Graduate Texts in Operations Research

This series contains compact volumes on the mathematical foundations of Operations Research, in particular in the areas of continuous, discrete and stochastic optimization. Inspired by the PhD course program of the Dutch Network on the Mathematics of Operations Research (LNMB), and by similar initiatives in other territories, the volumes in this series offer an overview of mathematical methods for post-master students and researchers in Operations Research. Books in the series are based on the established theoretical foundations in the discipline, teach the needed practical techniques and provide illustrative examples and applications.

Éric D. Taillard

Design of Heuristic Algorithms for Hard Optimization

With Python Codes for the Traveling Salesman Problem

Springer

Éric D. Taillard
University of Applied Sciences of Western
Switzerland, HEIG-VD
Yverdon-les-Bains, Switzerland

The open access publication of this book has been published with the support of the Swiss National Science Foundation.

ISSN 2662-6012 ISSN 2662-6020 (electronic)
Graduate Texts in Operations Research
ISBN 978-3-031-13713-6 ISBN 978-3-031-13714-3 (eBook)
https://doi.org/10.1007/978-3-031-13714-3

This Springer imprint is published by the registered company Springer Nature Switzerland AG
The registered company address is: Gewerbestrasse 11, 6330 Cham, Switzerland

Preface

This book is intended for students, teachers, engineers and scientists wishing to become familiar with metaheuristics. Frequently, metaheuristics are seen as an iterative master process guiding and modifying the operations of subordinate heuristics. As a result, the works in the field are organized into chapters, each presenting a metaheuristic, such as simulated annealing, tabu search, artificial ant colonies or genetic algorithms, to name only the best known.

This book addresses metaheuristics from a new angle. It presents them as a set of basic principles that are combined with each other to design a heuristic algorithm.

Heuristics and Metaheuristics

When addressing a new problem, we try to solve it by exploiting the knowledge acquired by experience. If the problem seems peculiarly difficult, a solution that is not necessarily the best possible is accepted. The matter is to discover the solution with a reasonable computational effort. Such a resolution method is then called a heuristic.

By analysing a whole menagerie of metaheuristics proposed in the literature, we identified five major basic principles leading to the design of a new algorithm:

1. **Problem modeling** The most delicate phase when confronted with a new problem is its modeling. Indeed, if a problem is taken by the "wrong end," its resolution can be largely compromised. Naturally, this phase is not the prerogative of metaheuristics.
2. **Decomposition into sub-problem** When one has to solve a complex problem or an instance of large size, it is necessary to decompose it into simpler or smaller sub-problems. These may themselves be difficult. Hence, they must be approached by an appropriate technique, for example a metaheuristic.
3. **Building a solution** When a suitable model is found, it becomes easy to build a solution to the problem, even if it is not good or even inapplicable in practice.

One of the most common construction methods is a greedy algorithm, which may even provide exact solutions for simple problems such as the minimum spanning tree or the shortest path.
4. **Modifying a solution** The next step tries to improve a solution by applying slight modifications. This approach can be seen as a translation to the discrete world of gradient methods for differentiable optimization.
5. **Randomization and learning** Finally, the repetition of constructions or modifications makes it possible to improve the quality of the solutions produced, provided that a random component and/or a learning process are involved.

Table 1 Context of application of a heuristic method and a metaheuristic framework

	Heuristics	Metaheuristic
Area of application	A generic optimization problem	Combinatorial optimization
Knowledge to include	Specific to the problem	Heuristic optimization methods
Data to provide	Numerical values of a problem instance	A generic optimization problem
Result	A heuristic solution to the instance	A heuristic algorithm

Metaheuristics have become an essential tool to tackle difficult optimization problems, even if they have sometimes been decried, especially in the 1980s by people who opposed exact and heuristic methods. Since then, it has been realized that many exact methods embed several heuristic procedures and did guarantee optimality only with limited precision!

Book Structure

This book is divided into three parts. The first part recalls some basics of linear programming, graph theory and complexity theory, and presents some simple and intractable combinatorial optimization problems. The aim of this first part is to make the field intelligible to a reader with no particular knowledge about combinatorial optimization.

The second part deals with the fundamental building blocks of metaheuristics: the construction and improvement of solutions as well as the decomposition of a problem into sub-problems. Primitive metaheuristics assembling these ingredients and using them in iterative processes running without memory are also incorporated.

The third part presents more advanced metaheuristics exploiting forms of memory and learning that allow the development of more elaborate heuristics. The exploitation of a memory can be done in different forms. One can try learning how to build good solutions directly on the basis of statistics gathered from previous trials.

Another possibility is to try to exploit memories to move intelligently through the solution space. Finally, one can store a whole set of solutions and combine them.

The book concludes with a chapter providing some advice on designing heuristics, an appendix providing source code for testing various methods discussed in the book and solutions to the exercises given at the end of each chapter.

Chapter 1 "Elements of Graphs and Complexity Theory" Before considering developing a heuristic for a problem, it is necessary to ensure the problem is difficult and that there is not an efficient algorithm to solve it exactly. This chapter includes a very brief introduction to two techniques for modeling optimization problems, linear programming and graph theory. Formulating a problem as a linear program describes it formally and unambiguously. Once expressed in this form and if the problem data is not too large, automatic solvers can be used to solve it. Some solvers are indeed built into spreadsheets of Office suites. With a little luck, there is no need to design a heuristic!
Many combinatorial optimization problems can be "drawn" and thus represented intuitively and relatively naturally by a graph. This book is illustrated by numerous examples of problems from graph theory. The traveling salesman problem is most likely the best known and has served as a guideline in the writing of this book.
Some elements of complexity theory are also presented in this introductory chapter. This area deals with the classification of problems according to their difficulty. Some simple techniques are given to show a problem is intractable. This helps to justify why it is essential to turn to the design of heuristics.

Chapter 2 "A Short List of Combinatorial Optimization Problems" This chapter reviews a number of classical problems in combinatorial optimization. It illustrates the sometimes narrow boundary between an easy problem, for which an efficient algorithm is known, and an intractable problem which differs merely in a small detail that may seem trivial at first sight.
Likewise, it allows the reader who is not familiar with combinatorial optimization to discover a broad variety of problems in various domains: optimal paths, traveling salesman, vehicle routing, assignment, network flow, scheduling, clustering, etc.

Chapter 3 "Problem Modeling" This chapter begins by describing techniques for simplifying the treatment of constraints, notably by transforming the objective into a fitness function. Then, it gives a brief overview of multi-objective optimization. Finally, it provides some practical applications of classical combinatorial optimization problems. It gives examples of data transformations to deal with applications that are at first sight far from classical descriptions.

Chapter 4 "Constructive Methods" This chapter presents methods for constructing solutions, starting with a reminder of the separation and evaluation methods, widely used for the design of exact algorithms. Next, two basic methods are presented, random construction and greedy construction. The latter sequentially selects the elements to be added to a partial solution, never questioning the choices that have been made. This method can be improved by evaluating more deeply the consequences of choosing an element. The *beam search* and the *pilot method* are part of these. The construction of a solution constitutes the first step in the design of a heuristic.

Chapter 5 "Local Search" The further step is to improve an existing solution by searching for minor changes that improve it. Local searches constitute the backbone of most metaheuristics. These methods are based on the definition of a set of neighbor solutions, for any solution of the problem. The definition of this set naturally depends on the modeling of the problem. Depending on the latter, a naturally expressed neighborhood may be too small to lead to quality solutions or, on the contrary, too large, leading to prohibitive computational times. Various methods have been proposed to enlarge the neighborhood, such as *filter and fan* or *ejection chains*, or to reduce it, such as *granular search* or *candidate list*.

Chapter 6 "Decomposition Methods" In the process of developing a new algorithm, this chapter should logically have been placed after the one devoted to problem modeling. However, decomposition methods are only used when the size of the data to be processed is large. It is, therefore, an optional phase, which the reader can ignore before moving on to stochastic and learning methods. This is the reason why it is placed at the end of the first part of this book, devoted to the key ingredients of metaheuristics. In this chapter, we consider methods like *POPMUSIC* or more general methods such as *large neighborhood search* or *fix-and-optimize*.

Chapter 7 "Randomized Methods" This chapter is devoted to methods repeating randomly and without memory constructions or modifications of solutions. Among the most popular techniques, we find *GRASP*, which integrates two basic bricks of metaheuristics: a randomized greedy construction and a local search. Four randomized local searches are presented in this chapter, showing that with the same classic recipe, different heuristics can be obtained: *simulated annealing*, *threshold accepting*, *great deluge* and the *noising* methods. The *variable neighborhood search* equally finds its place in this chapter.

Chapter 8 "Construction Learning" Following the order in which the key ingredients of metaheuristics are presented, one can first seek to improve the solution building process. Having constructed many solutions, one can collect statistics on their structure and exploit this data to construct new solutions. Artificial *ant colonies* represent a typical example. Another technique, *vocabulary building*, is also discussed in this chapter.

Chapter 9 "Local Search Learning" If local searches constitute the backbone of metaheuristics, the *taboo search*, which seeks to learn how to iteratively modify a solution, can be considered as the master of metaheuristics. The term "metaheuristic" was coined by its inventor. This chapter will focus on the ingredients that can be considered as the basis of taboo search, namely the use of memories and solution exploration strategies. Other ingredients of taboo search proposed by its inventor, such as candidate lists, ejection chains or vocabulary building, find a more logical place in other chapters.

Chapter 10 "Population Management" When one has a population of solutions, one can try learning how to combine them and how to manage this population. The most popular method in this field is undoubtedly *genetic algorithms*. However, genetic algorithms are a less advanced metaheuristic than *scatter search*, which provides strategies for managing a population of solutions. *GRASP method with path relinking* shows how to design a simple heuristic integrating several basic bricks of metaheuristics, ranging from randomized construction to population exploitation through local searches. Ultimately, among the latest metaheuristics, we find *particle swarm* methods, which seem to be adapted to continuous optimization. It should be noted that the spreadsheets of Office suites directly integrate this type of heuristic methods among the proposed solvers.

Chapter 11 "Heuristics Design" The concluding chapter of the book dispenses some advice on designing heuristics. It returns to the difficulty that can be encountered when modeling the problem and gives an example of decomposition into a chain of sub-problems for easy handling. Next, it proposes an approach for the development of a heuristic. Finally, some techniques for the parameter tuning and comparing the efficiency of algorithms are reviewed.

Source Codes for the Traveling Salesman Problem

One of the reasons for the popularity of metaheuristics is that they allow addressing difficult problems with simple codes. This book contains several pieces of code illustrating how to implement the basic methods discussed. Due to a certain level of abstraction, these principles could be perceived as a sculpture on clouds. The codes eliminate all ambiguity on the inevitable interpretations that can be done when presenting a metaheuristic framework. The computer scientist wishing to develop a heuristic method for a particular problem can be inspired by these codes.

As a source code is useful only when one wants to know all the details of a method, but that it is of little interest when reading. So, we have simplified and shortened the codes, trying to limit them to a single page. These codes come in addition to the text and not the opposite. The reader with little interest in programming or not familiar with the programming language used can skip them.

These codes have been tested and are relatively efficient. They addressed the emblematic traveling salesman problem. The latter is pedagogically interesting because its solution can be graphically drawn. Certainly, these codes are not "horse race," but they contain the quintessence of methods, and their extreme brevity should allow the reader to understand them. More than a dozen different methods have been implemented, while jointly taking less than one-tenth of the number of lines of code of one of the fastest implementations. However, we had to comply with this brevity. The codes are somewhat compact, succinctly commented and sometimes placing several instructions on the same line. So, we ask the reader used to sparser codes to be indulgent.

Exercises

Many exercises have been imagined. Their choice has always been guided by the solutions expected, which must be as unambiguous as possible. Indeed, when designing a new heuristic, there is no more either a correct or a false solution. There are only heuristics that work well for some problem instances and others which do not produce satisfactory results—bad quality of solutions, prohibitive calculation time, etc.

Nothing is more destabilizing for a student than ambiguous or even philosophical responses. This reason has led us to provide the solutions to all the exercises.

Acknowledgements

This book was published with the support of the Swiss National Science Foundation, grant number 10BP12_211862. It is the result of several years of teaching metaheuristics at Swiss universities of applied sciences and at the Universities of Vienna, Nantes, Hamburg and Graz. It was partially written during a sabbatical granted by the HEIG-VD. Unfortunately, I cannot name individually all the friends, colleagues and students who have contributed more or less directly and inspired me. I am thinking in particular of the many discussions and pleasant moments spent with my colleagues and friends during conferences, notably the successive organizers of the Metaheuristic International Conference. Nevertheless, I would like to thank especially some people who helped me to improve this book during the last months: Ante Rosandic, who translated a first version of the codes from C to Python, Peter Grestorfer, Hassan Taheri, and anonymous experts for their suggestions and corrections.

Yverdon-les-Bains, Switzerland — Éric D. Taillard
June 2022

Contents

Part I
Combinatorial Optimization, Complexity Theory, and Problem Modeling

This part is a gentle introduction to some basics of linear programming, graph theory, and complexity theory and presents some simple and difficult combinatorial optimization problems. The purpose of this introductory part is to make the domain intelligible to a reader who does not have specific knowledge in modeling such problems.

Chapter 1
Elements of Graphs and Complexity Theory

Before designing a heuristic method to find good solutions to a problem, it is necessary to be able to formalize it mathematically and to check that it belongs to a difficult class. Thus, this chapter recalls some elements and definitions in graph theory and complexity theory in order to make the book self-contained. On the one hand, basic algorithmic courses very often include graph algorithms. Some of these algorithms have simply been transposed to solve difficult optimization problems in a heuristic way. On the other hand, it is important to be able to determine whether a problem falls into the category of difficult problems. Indeed, one will not develop a heuristic algorithm if there is an efficient algorithm to find an exact solution.

1.1 Combinatorial Optimization

The typical field of application of metaheuristics is combinatorial optimization. Let us briefly introduce this domain with an example of a combinatorial problem: the coloring of a geography map. It is desired to assign a color for each country drawn on a map so that any two countries that have a common border do not receive the same color. In Fig. 1.1, five different colors are used, without worrying about the political attribution of the islands or enclaves.

This is a combinatorial problem. Indeed, if there are n areas to color with five colors, there are 5^n different ways to color the map. Most of these colorings are unfeasible because they do not respect the constraint that two areas with a common border do not receive an identical color. The question could be asked whether there is a feasible coloring using only four colors. More generally, one may want to find a coloring using a minimum number of colors. Consequently, we are dealing here with a *combinatorial optimization problem*.

É. D. Taillard, *Design of Heuristic Algorithms for Hard Optimization*, Graduate Texts in Operations Research, https://doi.org/10.1007/978-3-031-13714-3_1

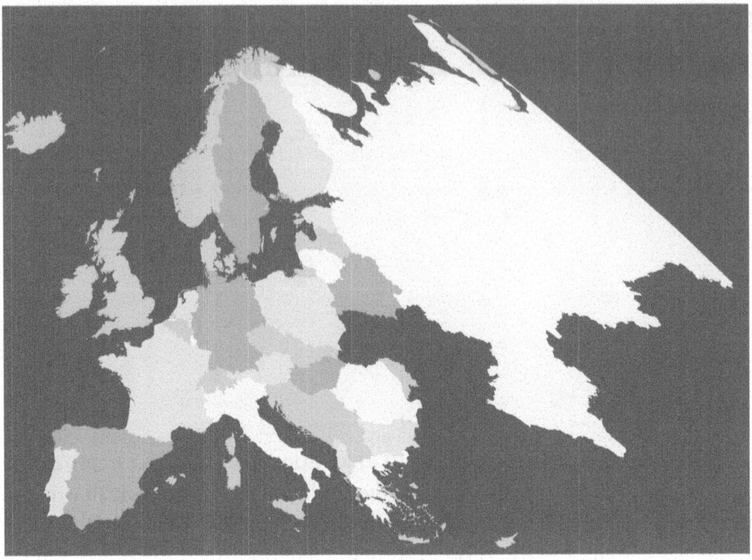

Fig. 1.1 An old European map colored with five colors (taking the background into account)

How to model this problem more formally? Let us take a smaller example (see Fig. 1.2): can we color Switzerland (s) and neighboring countries (Germany d, France f, Italy, Liechtenstein, and Austria a) with three colors?

A first model can be written using 18 binary variables that are put into equations or inequations. Let us introduce variables $x_{s1}, x_{s2}, x_{s3}, x_{d1}, x_{d2}, \dots, x_{a3}$ that should either take value 1 or 0. $x_{ik} = 1$ means that country i receives color k. Now, we can impose that a given country i receives exactly one color by writing the equation: $x_{i1} + x_{i2} + x_{i3} = 1$. To avoid assigning the same color to two countries (i and j) having a common border, we can write three inequalities (one for each color): $x_{i1} + x_{j1} \leqslant 1$, $x_{i2} + x_{j2} \leqslant 1$ and $x_{i3} + x_{j3} \leqslant 1$.

Another model can introduce 18 Boolean variables $b_{s1}, b_{s2}, b_{s3}, b_{d1}, b_{d2}, \dots, b_{a3}$ that indicate the color (1, 2 or 3) of each country. $b_{ik} = true$ means that country i receives color k. Now, we write a long Boolean formula that is true if and only if there is a feasible 3-coloring. First of all, we can impose that Switzerland is colored with at least one color: $b_{s1} \vee b_{s2} \vee b_{s3}$. But it should not receive both color 1 and color 2 at the same time: This can be written $\overline{b_{s1} \wedge b_{s2}}$, which is equivalent to $\overline{b_{s1}} \vee \overline{b_{s2}}$. Then, it should also not receive both color 1 and color 3 or color 2 and color 3. Thus, to impose that Switzerland is colored with exactly 1 color, we have the conjunction of four clauses:

$$(b_{s1} \vee b_{s2} \vee b_{s3}) \wedge (\overline{b_{s1}} \vee \overline{b_{s2}}) \wedge (\overline{b_{s1}} \vee \overline{b_{s3}}) \wedge (\overline{b_{s2}} \vee \overline{b_{s3}})$$

For each of the countries concerned, it is also necessary to write a conjunction of four clauses but with the variables corresponding to the other countries. Finally, for each border, it is necessary to impose that the colors on both sides are different. For example, for the border between Switzerland and France, we must have:

$$(\overline{b_{s1}} \vee \overline{b_{f1}}) \wedge (\overline{b_{s2}} \vee \overline{b_{f2}}) \wedge (\overline{b_{s3}} \vee \overline{b_{f3}})$$

Now a question arises: how many variables are needed to color a map with n countries which have a total of m common borders using k colors? Another one is: how many constraints (equation, inequation, or clauses) are needed to describe the problem? First, it is necessary to introduce $n \cdot k$ variables. Then, for each country, we can write one equation or $1 + \frac{k \cdot (k-1)}{2}$ clauses to be sure that each country receives exactly one color. Finally, for each border, it is necessary to write one inequation or $m \cdot k$ clauses. The problem of coloring such a map with k colors has a solution if and only if there is a value 1 or 0 for each of the binary variables or a value $true$ or $false$ for each of the Boolean variables such that all the constraints are simultaneously satisfied.

The Boolean model is called the *satisfiability* problem (SAT). It plays a central role in complexity theory. This extensive development is to formalize the problem by a set of equations or inequations or by a unique, long Boolean formula, but does not inform us yet how to discover a solution!

An extremely primitive algorithm to find a solution is to examine all the possible values for the variables (there are 2^{nk} different sets of values), and for each set, we have to check if the formula is true.

As modeled above, coloring a map is a *decision* problem. Its solution is either $true$ (a feasible coloring with k colors exists) or $false$ (this is impossible). Assuming that an algorithm $\mathscr{A}$ is available to obtain the values to assign to the variables so that all equations or inequations are satisfied or the Boolean formula is true—or to say that such values do not exist—is it possible to solve the *optimization* problem: which is the minimum number k of colors for having a feasible coloring?

One way to answer this question is to note that we need at most n colors for n areas and to assign a distinct color to each of them. As a result, we know that an n-coloring exists. We can apply the algorithm $\mathscr{A}$ to ask for an $n-1$ coloring, then $n-2$, etc. until getting the answer that no coloring exists. The ultimate value for which the algorithm has found a solution corresponds to an optimal coloring.

A faster technique is to proceed by a dichotomy: rather than reducing the number of color by one unit at each call of algorithm $\mathscr{A}$, two values, k_{min} and k_{max}, are stored so that it is known that there is no feasible coloring (respectively, a feasible coloring exists). By eliminating the case of the trivial map that has no boundary, we know that we can start with $k_{min} = 1$ and $k_{max} = n$. The algorithm is asked for a $k = \lfloor \frac{k_{min}+k_{max}}{2} \rfloor$ coloring. If the answer is yes, we modify $k_{max} \leftarrow \lfloor \frac{k_{min}+k_{max}}{2} \rfloor$; if the answer is no, we change $k_{min} \leftarrow \lfloor \frac{k_{min}+k_{max}}{2} \rfloor$. The method is repeated until $k_{max} = k_{min} + 1$. This value corresponds to the optimum number of colors. So, an optimization problem can be solved with an algorithm answering the corresponding decision problem.

1.1.1 Linear Programming

Linear programming is an extremely useful tool for mathematically modeling many optimization problems. *Mathematical programming* is the selection of a best element, with regard to some quantitative criterion, from some set of feasible alternatives. When the expression of this criterion is a linear function and all feasible alternatives can be described by means of linear functions, we are talking about linear programming.

A *linear program under canonical form* can be mathematically written as follows:

$$\text{Maximize} \quad z = c_1x_1 + c_2x_2 + \cdots + c_nx_n \tag{1.1}$$

$$\begin{aligned} \text{Subject to:} \quad & a_{11}x_1 + a_{12}x_2 + \cdots + a_{1n}x_n \leqslant b_1 \\ & a_{21}x_1 + a_{22}x_2 + \cdots + a_{2n}x_n \leqslant b_2 \\ & \dots \qquad\qquad \dots \qquad\qquad \dots \\ & a_{m1}x_1 + a_{m2}x_2 + \cdots + a_{mn}x_n \leqslant b_m \end{aligned} \tag{1.2}$$

$$x_j \geqslant 0 \ (j = 1, \dots, n) \tag{1.3}$$

z represents the *objective function* and x_j the *decision variables*. For a production problem, the c_j can be seen as revenues, the b_i being quantities of raw material available and the a_{ij} the unit consumption of material i for the production of good j.

The canonical form of linear programming is not limiting, in the sense that any linear program can be expressed under this form. Indeed, if the objective is to minimize z, this is equivalent to maximizing $-z$; if a variable x can be either positive or negative or null, it can be substituted by $x'' - x'$, where x'' and x' must be nonnegative; finally, if we have an equality constraint $a_{i1}x_1 + a_{i2}x_2 + \cdots + a_{in}x_n = b_i$, it can be replaced by the constraints $a_{i1}x_1 + a_{i2}x_2 + \cdots + a_{in}x_n \leqslant b_i$ and $-a_{i1}x_1 - a_{i2}x_2 - \cdots - a_{in}x_n \leqslant -b_i$.

The map coloring problem can be modeled by a slightly special linear program. For that, one introduces the variables y_k that indicate if the color k is used ($y_k = 1$) or not ($y_k = 0, k = 1, \dots, n$) in addition to the variables x_{ik} that indicate if the area i receives the color k. The *integer linear program* allows formalizing the optimization version of the map coloring problem:

$$\text{Minimize} \ z = \sum_{k=1}^{n} y_k \tag{1.4}$$

Subject to:

$$\sum_{k=1}^{n} x_{ik} = 1 \quad i = 1, \dots, n \tag{1.5}$$

$$x_{ik} - y_k \leqslant 0 \quad i, k = 1, \dots, n \tag{1.6}$$

$$x_{ik} + x_{jk} \leqslant 1 \quad \forall (i, j) \text{ having a common border}, \quad k = 1, \dots, n \tag{1.7}$$

$$x_{ik}, y_k \in \{0, 1\} \tag{1.8}$$

The objective (1.4) is to use the minimum number of colors. The first set of constraints (1.5) imposes that each vertex receives exactly one color; the second set (1.6) ensures that a vertex is not assigned to an unused color; the set (1.7) prevents the same color to be assigned to contiguous areas. The integrity constraints (1.8) can also be written with linear inequalities ($y_k \geqslant 0, \quad y_k \leqslant 1, \quad y_k \in \mathbb{Z}$).

Linear programming is a very powerful tool for modeling and formalizing problems. If there are no integrity constraints, problems with thousands of variables and thousands of constraints can be effectively solved. In this case, the resolution is barely more complex than the resolution of a system of linear equations. The key limitation is essentially due to the memory space required for data storage as well as any numerical problems that may occur if the data is poorly conditioned.

However, integer linear programs, like the coloring problem expressed above, are generally difficult to solve, and specific techniques should be designed. Metaheuristics are among these techniques.

If the formulation of a problem under the form of a linear program allows a rigorous modeling, it does not help our mind much for its solving. Indeed, the sight is the most important of our senses. The adage says a small drawing is better than a long speech. The graphs represent a more appropriate way for our spirit to perceive a problem. Before presenting other models for the coloring problem (see Sect. 2.8), some definitions in graph theory are recalled so that this book is self-contained.

1.1.2 A Small Glossary on Graphs and Networks

Graphs are a very useful tool for problem modeling when there are elements that have relationships between them. The elements are represented by a point and two related elements are connected by a segment. Thus, the previously seen map coloring problem can be drawn by a small graph, as shown in the Fig. 1.2.

1.1.2.1 Undirected Graph, Vertex, (Undirected) Edge

An *undirected graph* G is a pair of a set V of elements called *vertices* or *nodes* and of a set E of *undirected edges*, each of them associated with a (unordered) pair of

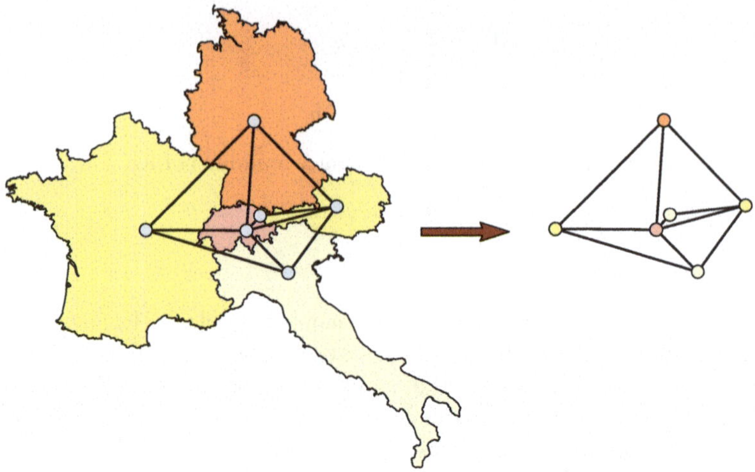

Fig. 1.2 Switzerland and its neighbor countries that we want to color. Each country is symbolized by a disk, and a common border is symbolized by a line connecting the corresponding countries. The map coloring can be transformed into the coloring of the vertices of a graph

nodes, which are their *endpoints*. Such a graph is noted as $G = (V, E)$. A vertex of a graph is represented by a point or a circle. An edge is represented by a line.

If two vertices v and w are joined by an edge, they are *adjacent*. The edge is *incident* with v and w.

When several edges connect the same pair of vertices, we have *multiple* edges. When both endpoints of an edge are the same vertex, this is a *loop*.

When $V = \varnothing$ (and $E = \varnothing$), we have the *null graph*. When $V \neq \varnothing$ and $E = \varnothing$, we have an *empty graph*. A graph with no loop and no multiple edges is a *simple graph*; otherwise, this is a *multigraph*. Figure 1.2 depicts a simple graph.

The complement graph $\overline{G}$ of a simple graph G has the same set of vertices and two distinct vertices of $\overline{G}$ are adjacent if and only if they are not adjacent in G.

1.1.2.2 Directed Graph, Arcs

In some cases, the relationships between the pairs of elements are ordered. This is a *directed graph* or *digraph*. The edges of a digraph are called the *arcs* or *directed edges*. An arc is represented by an arrow connecting its endpoints.

It is therefore necessary to distinguish both endpoints of an arc (i, j). The starting point i is called the *tail* and the arrival point j is the *head*. j is a *direct successor* of i and i is a *direct predecessor* of j. The set of direct successors of a node i is written $Succ(i)$, and the set of its direct predecessors $Pred(i)$.

An arc whose tail and head are the same vertex is also called a loop, as for the undirected case. Two arcs having the same tail and the same head are *parallel* or *multiple* arcs.

1.1.2.3 Incidence Matrix

The *incidence matrix* **A** of a graph with n vertices and m arcs and without loops is a matrix with m columns and n rows. The coefficients $a_{ij} (i = 1, \ldots, n, j = 1, \ldots, m)$ of **A** are defined as follows:

$$a_{ij} = \begin{cases} -1 \text{ if } i \text{ is the tail of the arc } (i, j) \\ 1 \text{ if } j \text{ is the head of the arc } (i, j) \\ 0 \text{ else} \end{cases}$$

In the case of an undirected graph, both endpoints are represented by 1s in the vertex-edge incidence matrix. It should be noticed that the incidence matrix does not allow to properly represent loops.

1.1.2.4 Adjacency Matrix

The *adjacency matrix* of a simple undirected graph is a square matrix with the coefficient a_{ij} is 1 if vertices i and j are adjacent and 0 otherwise.

1.1.2.5 Degree

The *degree* of a vertex v of an undirected graph, noted $deg(v)$, is the number of edges that are incident to v. A loop increases by 2 the degree of a vertex. A vertex of degree 1 is *pendent*. A graph is *regular* if all its vertices have the same degree. For a directed graph, the *outdegree* of a vertex, noted $deg^+(v)$, is the number of arcs having v as tail. The *indegree* of a vertex, $deg^-(v)$, is the number of arcs having v as head.

1.1.2.6 Path, Simple Path, Elementary Path, and Cycle

A *path* (also referred to as a *walk*) is an alternating sequence of vertices and edges, beginning and ending with a vertex, such that each edge is surrounded by its endpoints. A *simple path* (also referred to as a *trail*) is a walk for which all edges are distinct. An *elementary path* (also simply referred to as a path) is a trail in which all vertices (and therefore also all edges) are distinct. A *cycle* is a trail where the first vertex is corresponding to the last vertex. A *simple cycle* is a cycle in which the only repeated vertex is the first/last one. The *length* of a walk is its number of edges. Contrary to French, there is no difference in the wording between undirected and directed graphs. So, the edges, paths, etc. must be qualified with "directed" or "undirected." However, arcs are always directed edges.

1.1.2.7 Connected Graph

An undirected graph is *connected* if there is a path between every pair of its vertices. A *connected component* of a graph is a maximal subset of its vertices (and incident edges) such that there is a path between every pair of the vertices. A directed graph is *strongly connected* if there is a directed path in both directions between any pair of vertices.

1.1.2.8 Tree, Subgraph, and Line Graph

A *tree* is a connected graph without cycles (*acyclic*). A *leaf* is a pendent vertex of a tree. A *forest* is a graph without cycles. Each of its connected component is a tree. A *rooted tree* is a directed graph with a unique path from one vertex (the root of the tree) to each remaining vertex.

$G' = (V', E')$ is a *subgraph* of $G = (V, E)$, if $V' \subset V$ and E' has all the edges of E with both endpoints in V'. A *spanning tree* of a graph G is a subgraph of G which is a tree.

The *line graph* $L(G)$ of a graph G is built as follows (see also Fig. 2.12):

- Each edge of G is associated with a vertex of $L(G)$.
- Two vertices of $L(G)$ are joined by an edge if their corresponding edges in G share an endpoint.

1.1.2.9 Eulerian, Hamiltonian Graph

A graph is *Eulerian* if it contains a walk that uses every edge exactly once. A graph is *Hamiltonian* if it contains a walk that uses every vertex exactly once. Sometimes, Eulerian and Hamiltonian graphs are limited to the case when there is a cycle that uses every edge or every vertex exactly once (the first/last vertex excepted).

1.1.2.10 Complete, Bipartite Graphs, Clique, and Stable Set

In a *complete graph*, every two vertices are adjacent. All edges that could exist are present. A *bipartite graph* $G = (V, E)$ is such that $V = V_1 \cup V_2$, $V_1 \cap V_2 = \varnothing$ and each edge of E has one endpoint in V_1 and the other in V_2. A *clique* is a maximal set of mutually adjacent vertices that induces a complete subgraph. A *stable set* or *independent set* is a subset of vertices that induces a subgraph without any edges. A number of elements defined in the above paragraphs are illustrated in Fig. 1.3

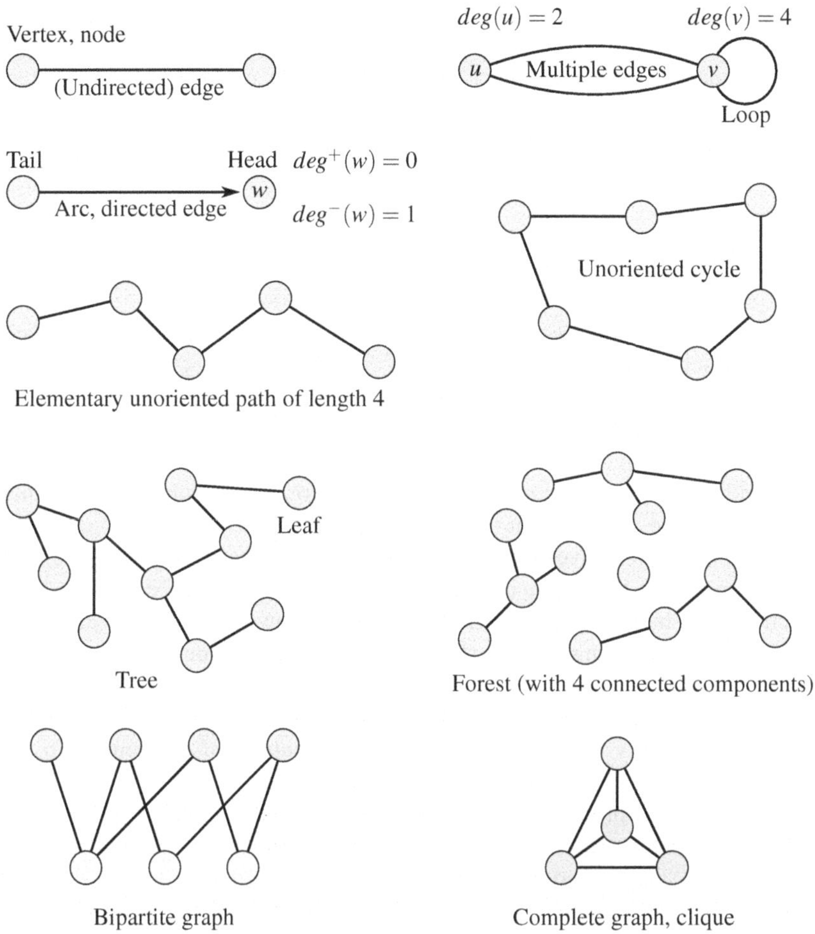

Fig. 1.3 Basic definition of graph components

1.1.2.11 Graph Coloring and Matching

The *vertex coloring* problem has been used as an introductory example in the Sect. 1.1 devoted to combinatorial optimization. A proper coloring is a labeling of the vertices of a graph by elements from a given set of colors such that distinct colors are assigned to the endpoints of each edge. The *chromatic index* of a graph G, noted $\chi(G)$, represents the minimum number of colors of a proper coloring of G. An *edge coloring* is a labeling of the edges by elements from a set of colors. The proper edge coloring problem is to minimize the number of colors required so that two incident edges do not receive the same color. A *matching* is a set of edges sharing no common endpoints. A *perfect matching* is a matching that matches every vertex of the graph.

1.1.2.12 Network

In many situations, a *weight* $w(e)$ is associated with every edge e of a graph. Typically, $w(e)$ is a distance, a capacity or a cost. A *network*, noted $R = (V, E, w)$, is a graph together with a function $w : E \to \mathbb{R}$. The *length* of a path in a network is the sum of the weights of its edges.

1.1.2.13 Flow

A classical problem in a directed network $R = (V, E, w)$ is to assign a nonnegative *flow* x_{ij} to each edge $e = (i, j)$ so that $\sum_{j \in Succ(i)} x_{ij} = \sum_{k \in Pred(i)} x_{ki} \quad \forall i \in V, i \neq s, t$. Vertex s is the source-node and t the sink-node. If $0 \leqslant x_{ij} \leqslant w_{ij} \forall (i, j) \in E$, the flow from s to t is *feasible*.

A *cut* is a partition of the vertices of a network $R = (V, E, w)$ into two subsets $A \subset V$ and $\overline{A} \subset V$. The *capacity* of a cut from A to $\overline{A}$ is the sum of the weight of the edges that have one endpoint in A and the other in $\overline{A}$.

Network flows are convenient to model problems that have, at first glance, nothing in common with flows, like resource allocation problems (see, e.g., Sect. 2.5.1). Further, in this chapter, we will review some well-known and effective algorithms for the minimum spanning tree, the shortest path, or the optimum flow in a network. Other problems, like graph coloring, are intractable. The only algorithms known to solve them require a time that can grow exponentially with the size of the graph.

Complexity theory focuses on classifying computational problems into easy and intractable ones. Metaheuristics have been designed to identify satisfactory solutions to difficult problems, while requiring a limited computing effort. Before developing a new algorithm on the basis of the principles of metaheuristics, it is essential to be sure the problem addressed is an intractable one and that there are not already effective algorithms to solve it. The rest of this chapter exposes some theoretical bases in the field of classification of problems according to their difficulty.

1.2 Elements of Complexity Theory

The purpose of complexity theory is to classify the problems in order to predict whether they will be easy to solve. To limit ourselves to sequential algorithms, we consider, very roughly, that an *easy problem* can be solved by an algorithm, which computational effort is limited by a function that polynomially depends on the size of the data to be treated. We can immediately dare why the difficulty limit must be on the class of polynomials and not on that of logarithmic, trigonometric, or exponential functions.

The reason is very simple: we can perfectly conceive that more effort is required to process a larger volume of data, eliminating nongrowing functions like trigonometric ones. Limited to sequential methods, it is clear that each record must be read at least once, which implies a growth in the number of operations at least linear. This eliminates logarithmic, square root, etc. functions. Naturally, for a parallel treatment of the data by several tasks, it is quite reasonable to define a class of problems (very easy), requesting a number of operations and memory *per processor* increasing at most logarithmically with the data volume. An example of such a problem is finding the largest number of a set.

Finally, we must consider that an exponential function (in the mathematical sense, such as 2^x, but also extensions such as $x^{\log x}$, $x!$ or x^x) always grow faster than any polynomial. This growth is incredibly impressive.

Let us examine the example of an algorithm that requires 3^{50} operations for a problem with 50 elements. If this algorithm is run on a machine able to perform 10^9 operations per second, the machine will not complete its work before 23 million years. By comparison, solving a problem with ten elements—five times smaller—with the same algorithm would take only 60 microseconds.

Hence, it would not be reasonable in practice to consider as easy a problem requiring an exponential number of operations to be solved. But combinatorial problems include an exponential number of solutions. As a result, complete enumeration algorithms, sometimes called "brute force," cannot be reasonably considered acceptable. Thus, the computation of a shortest path between two vertices of a network cannot be solved by enumerating the complete set of all paths since it is exponentially large. Algorithms using mathematical properties of the shortest walks must be used. These algorithms perform a number of steps that is polynomial in the network size. On the one hand, finding a shortest walk is an easy problem. On the other hand, finding a longest (or a shortest) path (without circuits or without visiting twice the same vertex) between two vertices is an *intractable problem*, because no polynomial algorithm is known to solve it.

Finally, we must mention that the class of polynomials has an interesting property: it is closed. The composition of two polynomials is also a polynomial. In the context of programming, it means that a polynomial number of calls to a subroutine that requires a computational effort that grows polynomially with the data size leads to a polynomial algorithm.

1.2.1 Algorithmic Complexity

Complexity theory and algorithmic complexity should not be mixed up. As already mentioned, complexity theory focuses on the problem classification. The purpose of algorithmic complexity is to evaluate the resources required to run a given algorithm. It is therefore possible to develop an algorithm of high complexity for a problem belonging to the class of "simple" problems.

To be able to put a problem into a complexity class, we will not assume the use of any given algorithm to solve this problem, but we will analyze the performance of the best possible algorithm—not necessarily known—for this problem and running on a given type of machine. We must not confuse the simplicity of an algorithm (expressed, e.g., by the number of lines of code needed to implement it) and its complexity. Indeed, a naive algorithm can be of high algorithmic complexity.

For instance, to test if an integer p is prime, we can try to divide it by all the integers between 2 and $\sqrt{p}$. If all these divisions have a reminder, we can conclude that p is prime. Otherwise, there is a certificate (a divider of p) proving that p is not prime. This algorithm is easy to implement. However, it is not polynomial in the size of the data. Indeed, just $n = \log_2(p)$ bits are required to code the number p. Therefore, the algorithm requires a number of divisions proportional to $2^{n/2}$, which is not polynomial.

However, it has been proven in 2002 that there is a polynomial algorithm to detect if a number p is prime. As we can expect, this algorithm is undoubtedly a sophisticated one. Its analysis and implementation is just a task at the limits of human capacities. So, testing whether a number is prime or not remains a simple problem (because there is a polynomial algorithm to solve it). However, this algorithm is difficult to implement and would require a prohibitive computational time to prove that $2^{82,589,933} - 1$ is prime. Conversely, there are algorithms that could theoretically degenerate but that consistently behave appropriately in practice, like the simplex algorithm for linear programming.

The resources required during the execution of an algorithm are limited. They are of several types: number of processors, memory space, and time. Looking at this last resource, we could measure the effectiveness of an algorithm by evaluating its running time on a given machine. Unluckily, this measure presents many weaknesses. First, it is relative to a particular machine, whose lifetime is limited to a few years. Then, the way the algorithm has been implemented (programming language, compiler, options, operating system) can notably influence its running time. Therefore, it is preferred to measure the characteristic number of operations that an algorithm will perform. Indeed, this number does not depend on the machine or language and can be perfectly theoretically evaluated.

We call *complexity of an algorithm* a function $f(n)$ that gives the characteristic number of steps executed *in the worst case*, when it runs on a problem whose data size is n. It should be mentioned that this complexity has nothing to do with the length of the code or with the difficulty to code it. The average number of steps is also seldom used since this number is generally difficult to evaluate. Indeed, it would be necessary to take an average for all possible data sets. In addition, the worst-case evaluation is essential for applications where the running time is critical.

1.2.2 Bachmann-Landau Notation

In practice, a rough overestimate is used to evaluate the number of steps performed by an algorithm to solve a problem of size n. Suppose that two algorithms, $\mathscr{A}_1$ and $\mathscr{A}_2$ perform, respectively, for the same problem of size n, $f(n) = 10n^2$ and $g(n) = 0.2 \cdot n^3$ operations.

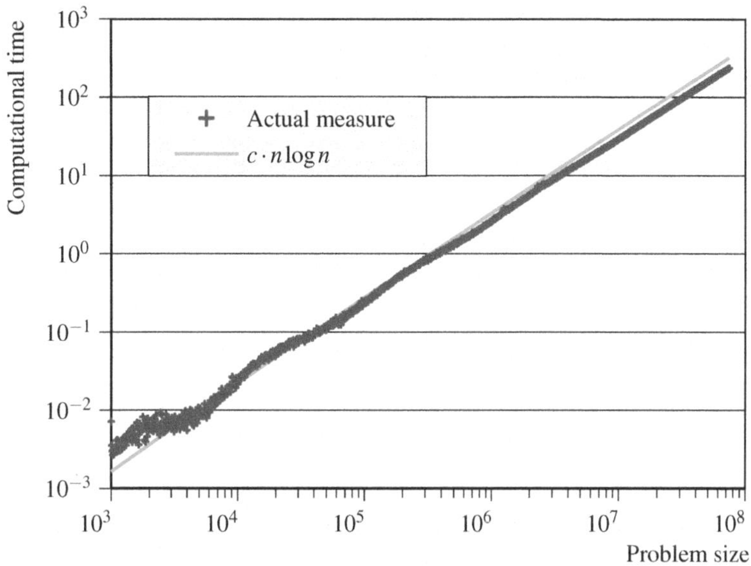

Fig. 1.4 Observed computational time for building a traveling salesman tour as a function of the number n of cities. For instances with more than a million cities, the time remains below the $c \cdot n \log n$ function. This verifies that the method is in $O(n \log n)$

On the one hand, for $n = 10$, it is clear that $\mathscr{A}_1$ performs five times more operations than $\mathscr{A}_2$. On the other hand, as soon as $n \geqslant 50$, $\mathscr{A}_2$ will perform more steps than $\mathscr{A}_1$.

As n grows large, the n^3 term will come to dominate. The positive coefficients in front of n^2 and n^3 in $f(n)$ and $g(n)$ become irrelevant. The function $g(n)$ will exceed $f(n)$ once n grows larger than a given value. The *order* of a function captures the asymptotic growth of a function.

1.2.2.1 Definitions

If f and g are two real functions of a real (or integer) variable n, it is said that f is of an order lower or equal to g if there are two positive constants n_0 and c such that $\forall n \geqslant n_0, f(n) \leqslant c \cdot g(n)$. This means that $g(n)$ grows larger than $f(n)$ as soon as $n \geqslant n_0$, irrespective of the constant factor c. With Bachmann-Landau notation, this is written $f(n) = O(g(n))$ or $f(n) \in O(g(n))$. This is the *big O* notation.

The diagram in Fig. 1.4 illustrates the usefulness of this notation. It gives the observed computation time to construct a traveling salesman's tour for various problem sizes. Observing the measurement dispersion for small sizes, it seems difficult to find a function for expressing the exact computational time. However, the observations for large sizes show the $n \log n$ behavior of this method, presented in Sect. 6.3.2.

The practical interest of this notation is that it is often easy to find a function g that increases asymptotically faster than the exact function f which may be difficult to evaluate. So, if the number of steps of an algorithm is smaller than $g(n)$ for large values of n, it is said that the algorithm runs at worst in $O(g(n))$.

Sometimes, we are not interested in the worst case but in the best case. It is said that $f(n) \in \Omega(g(n))$ if $f(n)$ increases asymptotically faster than $g(n)$.

Mathematically, $f(n) \in \Omega(g(n))$ if $\forall n \geqslant n_0,\ f(n) \geqslant c \cdot g(x)$. This is equivalent to $g(n) = O(f(n))$. This notation is useful to show that an algorithm $\mathscr{A}$ is less efficient than another $\mathscr{B}$: at best, the last performs at least as many steps than $\mathscr{A}$. It can also be used to show that an algorithm $\mathscr{C}$ is optimal: at worst, $\mathscr{C}$ performs a number of steps that is not larger than the minimum number of steps required by any algorithm to solve the problem.

If the best and the worst case are the same, i.e., if $\exists c_2 > c_1 > 0$ such that $c_1 \cdot g(n) \leqslant f(n) \leqslant c_2 \cdot g(n)$, then it is written $f(n) \in \Theta(g(n))$.

The $\Theta(\cdot)$ notation should be distinguished from a notion (often not well-defined) of an average complexity. Indeed, taking the example of the *Quicksort* algorithm to sort n elements, we can say it is in $\Omega(n)$ and in $O(n^2)$. But this algorithm is not in $\Theta(n \log n)$, even if its average computational time is proportional to $n \log n$.

Indeed, it can be proven that the mathematical expectation of the computational time of *Quicksort* for a set of n elements randomly mixed up is proportional to $n \log n$. The notations $\overline{O}(\cdot)$ (theoretical expected value) and $\hat{O}(\cdot)$ (empirical average) are used later in this book. However, they are not frequently used in the literature. To use them properly, we must specify which data set is considered and the probability of occurrence of each problem instance, etc.

In mathematics and more seldom in computer sciences, there also exist the *small o* notations:

- $f(n) \in o(g(n))$ if $\lim_{n\to\infty} \frac{g(n)}{f(n)} > 0$
- $f(n) \in \omega(g(n))$ if $\lim_{n\to\infty} \frac{f(n)}{g(n)} > 0$
- $f(n) \sim g(n)$ if $\lim_{n\to\infty} \frac{f(n)}{g(n)} = 1$

There are many advantages to express the algorithmic complexity of an algorithm with the big O notation:

- $f(n) \in O(g(n))$ means that $g(n)$ is larger than the true complexity; this often allows to find a function $g(n)$ with an easy calculus while finding $f(n) \in \Theta(g(n))$ would have been much more difficult.
- $25n^3 = O(3n^3)$ and $3n^3 = O(25n^3)$, this means that two functions that differ solely from a constant factor have the same order; this allows to ignore the relative speed of computers; instead of writing $O(25n^3)$, we can write $O(n^3)$ which is equivalent and simpler.
- $3n^3 + 55n = O(n^3)$, this means that the lower order terms can be neglected; only the larger power has to be kept.

It is important to stress that the complexity of an algorithm is a theoretical concept, which is derived by reflection and calculations. This can be established

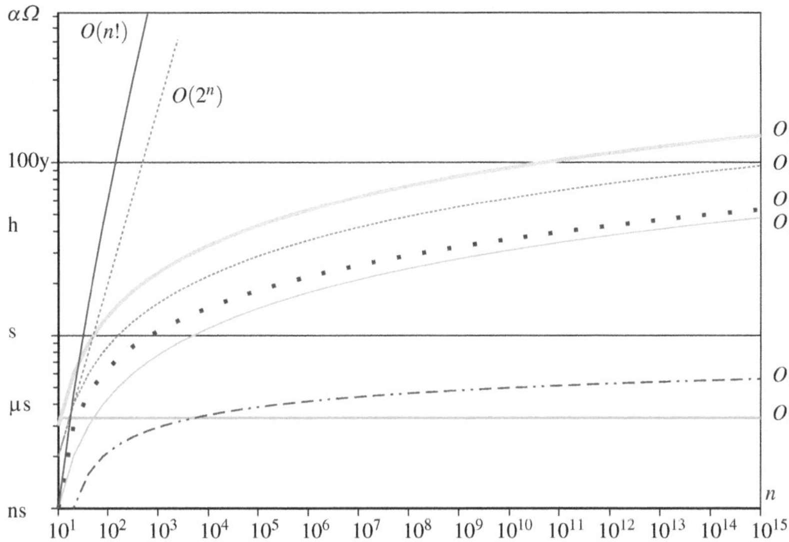

Fig. 1.5 Illustration of the growth of some functions frequently used to express the complexity of an algorithm. The horizontal axis indicates the size of the problem (with exponential growth) and the vertical axis gives the order of magnitude of the computation time (with iterated-exponential growth, from a nanosecond to the expected life of our universe)

with a sheet and a pencil. The complexity is typically expressed by the order of the computational time (or an abstract number of steps performed by a virtual processor) depending on the size of the problem.

Functions commonly encountered in algorithmic complexity are given below, with the slower-growing functions listed first. Figure 1.5 depicts the growth of some of these functions.

- $O(1)$: constant.
- $O(\log n)$: logarithmic; the base is not provided since $O(\log_a n) = O(\log_b n)$.
- $O(n^c)$: fractional power, with $0 < c < 1$.
- $O(n)$: linear.
- $O(n \log n)$: linearithmic.
- $O(n^2)$: quadratic.
- $O(n^3)$: cubic.
- $O(n^c)$: polynomial, with $c > 1$ constant.
- $O(n^{\log n})$: quasi-polynomial, super-polynomial, sub-exponential.
- $O(c^n)$: exponential, with $c > 1$ constant.
- $O(n!)$: factorial.

1.2.3 Basic Complexity Classes

Complexity theory has evolved considerably since the beginning of the 1970s, when Cook showed there is a problem which, if we were able to solve it in polynomial time, then it would allow us to solve many others efficiently, like the traveling salesman, the integer linear programming, the graph coloring, etc. [1].

To achieve this result, it was necessary to formulate a generic problem in mathematical terms, how a computer works, and how computational time can be measured. To simplify this theory as much as possible, the type of problems considered is limited to *decision* problems.

A decision problem is formalized by a generic problem and a question; the answer should be either "yes" or "no."

Example of a Generic Problem

Let $C = \{c_1, \ldots, c_n\}$ be a set of n cities, integer distances d_{ij} between the cities c_i and c_j $(i, j = 1, \ldots, n)$, and B an integer bound.

Question

Is there a tour of length not higher than B visiting every city of C? Put differently, we look for a permutation p of the elements $1, 2, \ldots, n$ such that $d_{p_n,p_1} + \sum_{i=1}^{n-1} d_{p_i}, d_{p_{i+1}} \leqslant B$.

This is the decision version of the *traveling salesman problem* (TSP for short). The optimization version of the problem seeks to find the shortest possible route that visits each city exactly once and returns to the origin city. This is undoubtedly the best-known combinatorial optimization problem that is intractable.

1.2.3.1 Encoding Scheme, Language, and Turing Machine

A problem instance can be represented as a text file. We must subsequently use given conventions, for example, put on the first line n, the number of cities, then B, the bound, on the second line, and each of the following line will contain three numbers, interpreted as i, j and d_{ij}. Put differently, an *encoding scheme* is used.

We can adopt the formal grammar of language theory, which is similar to those used in compiling techniques. Let Σ be a finite set of symbols or an alphabet. We write Σ^* the set of all strings that can be built with the alphabet Σ. An encoding scheme e for a generic problem π allows describing any instance I of π by a string $x \in \Sigma^*$. For the TSP, I contains n, B and all the d_{ij} values.

An encoding scheme e for generic problem π partitions the strings of Σ^* into three classes:

1. The strings that do not encode a problem instance I of π
2. The strings encoding a problem instance I of π for which the answer is "no"
3. The strings encoding a problem instance I of π for which the answer is "yes"

This last class is called the *language* associated with π and e, denoted $L(\pi, e)$.

In theoretical computer science, or more precisely in automata theory, the computing power of various machine models is studied. Among the simplest automata, there are finite-state automata. They are utilized to design or analyze a communication protocol for instance. Their states are represented by the vertices of a graph and transitions, represented by arcs. Providing an input string, the automaton changes from one state to another according to the symbol of the string being read and associated transitions rules. Since an automaton maintains a finite number of states, this machine possesses a bounded memory.

A slightly more complex model is a push-down automaton, functioning similarly to a finite-state machine, but has a stack. At each step, a symbol of the string is interpreted, as well as the symbol at the top of the stack (if the last is not empty). The automaton changes its state and places a new symbol at the top of the stack. This type of automaton is able to make more complex computations. For instance, it can recognize the strings of a non-contextual language. Hence, it can perform the syntax analysis of a program described by a grammar of type 2. An even more powerful computer model than a stack automaton is the Turing machine.

Deterministic Turing Machine

To mathematically represent how a computer works, Alan Turing imagined a fictive machine (there were no computers in 1936) whose operations can be modeled by a transition function. This machine is able to implement all the usual algorithms. It is able to recognize a string generated by a general grammar of type 0 in a finite time. Figure 1.6 illustrates such a machine, composed of a program that controls the scrolling of a magnetic tape and a read/write head.

A program for a deterministic Turing machine is specified by:

1. A tape alphabet Γ—the set of symbols that can be written on the tape. Γ contains at least Σ, the set of symbols that encodes a decision problem instance, the special blank symbol b not belonging to Σ and eventually other control symbols.
2. A set of states Q, containing at least q_0, the initial state, q_Y, the final state indicating that the answer to the instance is "yes" and q_N, the final state indicating that the answer is "no."
3. A transition function $\delta : Q \setminus \{q_Y, q_N\} \times \Gamma \quad \rightarrow \quad Q \times \Gamma \times \{-1, 1\}$.

This function represents the actions to be performed by the machine when it is in a certain state and reads a certain symbol. A Turing machine works as follows: its initial state is q_0, the read/write head is positioned on cell 1; the tape contains the

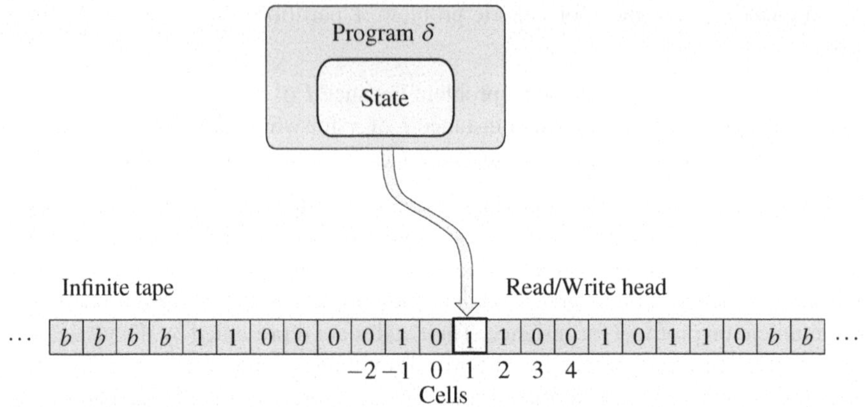

Fig. 1.6 Schematic representation of a deterministic Turing machine, which allows modeling and formalizing a computer

string $x \in \Sigma^*$ in cells 1 through $|x|$ and b for all other cells. Let q be the current state of the machine, σ the symbol read from the tape and $(q', \sigma', \Delta) = \delta(\sigma, q)$. One step of the machine consists in:

- Replacing σ by σ' in the current cell
- Moving the head one cell to the left if $\Delta = -1$ or one cell to the right if $\Delta = 1$
- Changing the internal state to q'

The machine stops either in state q_Y or in state q_N. This is the reason why the transition function δ is only defined for nonfinal states of the machine.

Although very simple, a Turing machine can conceptually represent everything that happens in a common computer. This is not the case for simpler machines, like the finite-state automaton (which head always moves toward the same direction) or the push-down automaton.

Example of a Turing Machine Program

Let $M = (\Gamma, \Sigma, Q, \delta)$ be a Turing Machine program:
Tape alphabet: $\Gamma = \{0, 1, b\}$
Input alphabet: $\Sigma = \{0, 1\}$
Set of states: $Q = \{q_0, q_1, q_2, q_3, q_Y, q_N\}$

Transition function δ: given in Table 1.1

Table 1.1 Specification of the transition function δ of a Turing machine

	Symbol $\sigma \in \Gamma$ on the tape		
State	0	1	b
q_0	$(q_0, 0, 1)$	$(q_0, 1, 1)$	$(q_1, b, -1)$
q_1	$(q_2, b, -1)$	$(q_3, b, -1)$	$(q_N, b, -1)$
q_2	$(q_Y, b, -1)$	$(q_N, b, -1)$	$(q_N, b, -1)$
q_3	$(q_N, b, -1)$	$(q_N, b, -1)$	$(q_N, b, -1)$

1.2.3.2 Class P of Languages

The class P (standing for *polynomial*) contains the problems considered easy: those for which an algorithm can solve the problem with a number of steps polynomially limited to the instance data size (the length of the string x initially written on the tape). More formally, this class is defined as follows: we say the machine M *accepts* $x \in \Sigma^*$ if and only if M stops in the state q_Y. The *language recognized* by M is the set of strings $x \in \Sigma^*$ such that M accepts x. We can verify that the language recognized by the machine given by the program in Table 1.1 is the strings encoding a binary number divisible by 4.

An *algorithm* is a program that stops for any string $x \in \Sigma^*$. The *computational time* of an algorithm is the number of steps performed by the machine before it stops. The *complexity* of a program M is the *largest* computational time $T_M(n)$ required by the machine to stop, whatever the string x of length n initially written on the tape is. A deterministic Turing machine program is in *polynomial time* if there is a polynomial p such that $T_M(n) \leqslant p(n)$

The *class* P of languages includes all the languages L such that there is a program for deterministic Turing machine recognizing L in polynomial time. By abuse of language, we say the problem π belongs to the class P if the language associated with π and with an encoding scheme e (unspecified but supposed to be reasonable) belongs to P. When we use the expression "there is a program"; we know this program exists, but without necessarily knowing how to code it. Conversely, if we are aware of an algorithm—not necessarily the best one—running in polynomial time for this problem, then the problem belongs to the complexity class P.

1.2.3.3 Class NP of Languages

Informally, the *class* NP (standing for *nondeterministic polynomial*) of languages includes all the problems for which we can *verify* in polynomial time that a given solution produces the answer "yes." For a problem to be part of this class, the requirements are looser than for the class P. Indeed, it is not required to be able to find a solution in polynomial time but only to be able to verify the correctness of a given solution in polynomial time. Practically, this class contains intractable problems, for which we are not aware of a polynomial time solving algorithm.

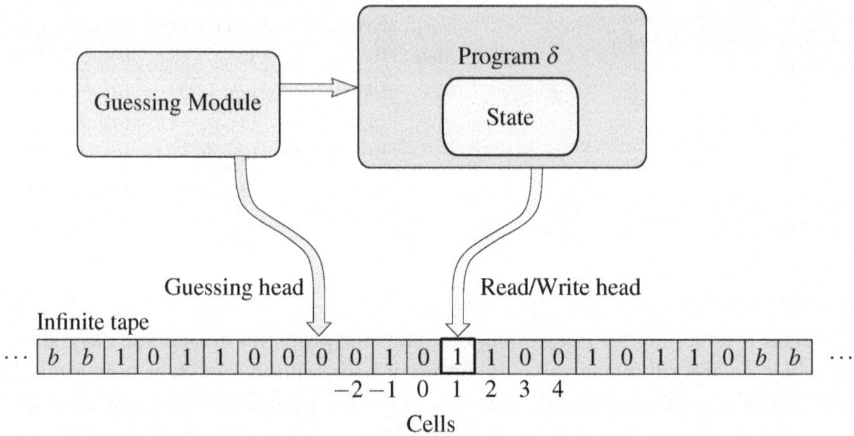

Fig. 1.7 Schematic representation of a nondeterministic Turing machine. This machine allows formalizing the NP class, but does not exist in the real world

To formalize this definition, theoretic computer scientists have imagined a new type of theoretical computer, the *nondeterministic Turing machine*, which has no material equivalent in our real world. Conceptually, this machine is composed of a module that *guesses* the solution of the problem and writes it into the negative index cells of the tape (see Fig. 1.7). This artifice allows us to overcome our ignorance of an efficient algorithm to solve the problem: the machine just does the job and guesses the solution.

The specification of a program for a nondeterministic Turing machine is identical to that of a deterministic one. Initially, the machine is in state q_0, the tape contains the string x encoding the problem in cells 1 to $|x|$, and the program is idle. At that time, a guessing phase starts during which the module writes random symbols in the negative cells and stops arbitrarily. Next, the machine's program is activated, and it works as a deterministic Turing machine.

With such a machine, it is obvious that a given string x can generate various computations, because of the nondeterministic character of the guessing phase. The machine can end in q_N state even if the problem includes a feasible solution. Different runs with various computational times can end in the q_Y state. But the machine cannot end in the state q_Y for a problem that has no solution.

By definition, the language L_M recognized by the nondeterministic machine M is the set of strings $x \in \Sigma^*$ such that there is *at least* one computation for which the string x is accepted. The computation time $T_M(n)$ is the *minimum* number of steps taken by the machine to accept a string x of length n. The number of steps in the guessing phase is not counted. The complexity of a program is defined in a similar way to that of a deterministic machine.

The *class NP of languages* is formally defined as the set of languages L for which there exists a program M for a nondeterministic Turing machine so that M recognizes L in *polynomial* time. We insist on the fact that the name of this class comes from "nondeterministic polynomial" and not from "non-polynomial."

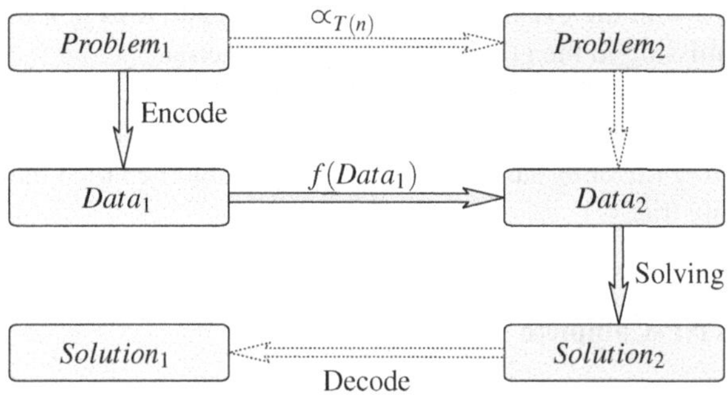

Fig. 1.8 Polynomial transformation of $Problem_1$ to $Problem_2$ in time $T(n)$. The theory only requires to be able to carry out the operations represented with solid line arrows

Polynomial Transformation

The notion of *polynomial transformation* of an initial problem into a second one is fundamental in the theory of complexity, because it is of substantial help for problem classification. Indeed, if we are able to efficiently solve the second problem—or, for intractable problems, if we were able to efficiently solve the second problem—and we know an inexpensive way of transforming the initial problem into the second one, then we can also effectively solve the initial problem.

Formally, a first language $L_1 \subset \Sigma_1^*$ can be *polynomially transformed* into a second language $L_2 \subset \Sigma_2^*$ if there is a function $f : \Sigma_1^* \to \Sigma_2^*$ that can be evaluated in polynomial time by a deterministic Turing machine, such that, for all problem instance $x \in \Sigma_1^*$ with "yes" answer, $f(x)$ is an instance of the second problem with "yes" answer. Such a polynomial transformation is written $L1 \propto L2$. We write $L_1 \propto_{T(n)} L_2$ if we want to specify the time $T(n)$ required to evaluate f.

Figure 1.8 illustrates the principle of a polynomial transformation. When transforming a problem into another one, it is solely concerned about the complexity of the evaluation of the f function and the answers "yes-no" of both instances should be the same. The complexity of solving instance 2 or that of the decoding of a solution of instance 1 from that of instance 2 is not required.

Example of Polynomial Transformation

Let us consider the problem of finding a Hamiltonian cycle in a graph (a cycle passing only once by all the vertices of the graph before returning to the starting vertex) and the traveling salesman problem. The last is to answer the question: is there a tour of total length no more than B? The f function to transform the Hamiltonian cycle into an instance of a traveling salesman builds a complete network on the same set of vertices as for the graph. In the network, it associates a weight of zero with the existing edges of the graph and a weight of one with the edges that are missing in the graph. The bound B is zero.

There is a solution of length 0 to the traveling salesman if and only if there is a Hamiltonian cycle in the initial graph. We deduce the Hamiltonian cycle can be transformed into a traveling salesman problem. It should be noted that the opposite is not necessarily true.

1.2.3.4 Class NP-Complete

A problem π belongs to the *class NP-complete* if π belongs to NP and every problem of NP can be polynomially transformed into π.

Starting from the definition of a polynomial transformation and noting the composition of two polynomials is still a polynomial, we have the following properties:

- If π is NP-complete and π can be solved in polynomial time, then $P = NP$.
- If π is NP-complete and π does not belong to P, then $P \neq NP$.
- If π_1 polynomially transforms into π_2 and π_2 polynomially transforms into π_3, then π_1 polynomially transforms into π_3.
- If π_1 is NP-complete, π_2 belongs to NP and π_1 polynomially transforms into π_2, then π_2 is NP-complete.

No NP-complete problem that can be solved in polynomial time is known. It is conjectured that no such problem exists, hence it is assumed that $P \neq NP$. The latter property listed above is frequently exploited to show that a problem π_2, of a priori unknown complexity, is NP-complete. For this, a problem π_1 belonging to the NP-complete class is chosen, and a polynomial transformation of any instance of π_1 into an instance of π_2 is exhibited.

The NP-complete class definition presented above is purely theoretical. Maybe, this class is just an empty one! Therefore, it should be asked whether there exists at least one problem belonging to this class or not? It is indeed far from obvious to find a "universal" problem of NP such that all the other problems of NP can be polynomially transformed into this problem. It is not possible to imagine what all the problems of NP are and even less to find a transformation for each of them into the universal problem. However, such a problem exists, and the first that was shown to be NP-complete was the satisfiability problem.

Satisfiability

Let $u_1, \ldots u_m$ be a set of Boolean variables. A *literal* is a variable or its negation. A (disjunctive) *clause* is a finite collection of literals connected together with logical "or" ($\vee$). A clause is false if and only if all its literals are false. A satisfiability problem is a collection of clauses connected together with the logical "and" ($\wedge$). An instance of satisfiability is feasible if there are assignments of values to the Boolean variables such that all the clauses are simultaneously true.

For instance, the satisfiability problem $(u_1 \vee \overline{u_2}) \wedge (\overline{u_1} \vee u_2)$ is feasible. However, $(u_1 \vee u_3) \wedge (u_1 \vee \overline{u_3}) \wedge (\overline{u_1}) \wedge (u_2)$ is not a feasible instance. The graph coloring problem modeled with a Boolean formula given at the very beginning of this chapter is a satisfiability problem.

In the early 1970s, Cook shows that satisfiability is NP-complete. From this result, it was quite easy to show that many others also belong to the class NP-complete, using the principle stated in the remark above. In the late 1970s, several hundred problems were shown to be NP-complete.

Below is the example of the polynomial transformation of satisfiability into the stable set problem. Since any problem of NP can be transformed into satisfiability and any satisfiability instance can be transformed into the stable set, the latter is NP-complete.

Stable Set

Data: a graph $G = (V, E)$ and k an integer. Question: Is there a subset $V' \subseteq V, |V'| = k$ such that $\forall i, j \in V', (i, j) \notin E$ (i.e., a subset of k nonadjacent vertices)?

Satisfiability is transformed into stable set as follows:

- A vertex is associated with all literals of each clauses.
- For each clause, a complete subgraph is created.
- Incompatible literals-vertices are connected together (a variable and its negation).
- A stable set of k vertices is searched in this graph, where k is the number of clauses.

Such a transformation is illustrated in Fig. 1.9 for a little instance with three literals and three clauses.

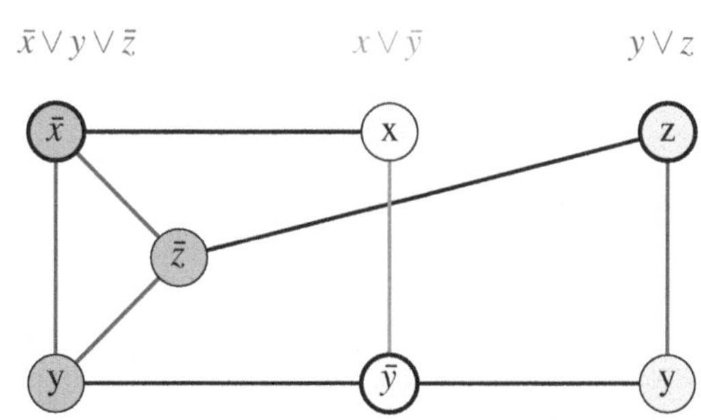

Fig. 1.9 Polynomial transformation of satisfiability instance: $(\bar{x} \vee y \vee \bar{z}) \wedge (x \vee \bar{y}) \wedge (y \vee z)$ to a stable set

Example of Unknown Complexity Problems

At this time, thousands of problems have been identified to be either in P or in NP-complete class. A number of them are not yet classified more precisely than in NP. Here are two examples of such problems:

- In a soccer league, each team play each other once. The winning team receives three points. The losing team receives zero points. In case of a tie, each team receives one point. Given a series of scores for each team, can this series be the result obtained at the end of a championship? Note: if the winner receives only two points, then there is a polynomial algorithm to answer this question.
- Is it possible to orient the edges of a graph so that it is strongly connected and that each vertex has an odd indegree?

1.2.3.5 Strongly NP-Complete Class

In some cases, NP-complete problem instances are well solved by means of ad hoc algorithms. For instance, dynamic programming can manage knapsack problems (see Sect. 2.5.3) with numerous items. A condition for these instances to be easily solved is that the largest number appearing in the data is limited. For the knapsack problem, this number is its volume. On the contrary, other problems cannot be solved effectively, even if the value of the largest number appearing in the problem is limited.

We are addressing a *number problem* if there is no polynomial $p(n)$ such that the largest number M appearing in the data of an instance of size n is bounded by $p(n)$. The partition of a set into two subsets of equal weight or the traveling salesman are, therefore, problems on numbers because, if we add one bit to the size of the problem, M can be multiplied by two. Therefore, for these problems, M can be in $O(2^n)$, which is not polynomial.

We say an algorithm is *pseudo-polynomial* if it runs in a time bounded by a polynomial depending on the size n of the data and the largest number M appearing in the problem. The partition of a set into two subsets of equal weight is an NP-complete problem for which there is a simple pseudo-polynomial algorithm.

Instance of a Partition Problem

Is it possible to divide the set $\{5, 2, 1, 6, 4\}$ into two subsets of equal weights? The sum of the weights for this partition problem instance is 18. Therefore, we look for two subsets of weight 9.

To solve this problem, we create an array of n rows, where n is the number of elements in the set, and $M = 9$ columns, where M is half of the sum of the element weights. We eventually fill the cells of this table with $\times$ by proceeding line by line. Using only the first element, of weight 5, we manage to create a subset of weight 0 (if we do not take this element) or a subset of weight 5 (taking it). Hence, we place $\times$ in the columns 0 and 5 of the first line.

Using only the first two elements, it is possible to create subsets whose weight is the same as with a single element (by not taking the second element). In the second line of the table, we can copy the $\times$ of the previous line. By taking the second element, we can create subsets of weights 2 and 7. Hence, we put $\times$ where we put them for the previous line but shifted by the weight of the second element (here: 2).

The process is then repeated until all the elements have been considered. As soon as there is a $\times$ in the last column, it means it is possible to create a subset of weight M. This is the case for this instance. One solution is {2, 1, 6}{5, 4}. The complexity of the algorithm is $O(M \cdot n)$, which is indeed polynomial in n and M.

	Sum of the weights									
Element	0	1	2	3	4	5	6	7	8	9
5	×					×				
2	×		×			×		×		
1	×	×	×	×		×	×	×	×	
6	×	×	×	×		×	×	×	×	×
4	×	×	×	×	×	×	×	×	×	×

Let π be a number problem and $\pi_{p(n)} \subset \pi$, the subset restricted to instances for which $M \leqslant p(n)$. The set $\pi_{p(n)}$ contains only instances of π with "small" numbers. It is said that π is *strongly NP-complete* if and only if there is a polynomial $p(n)$ such that $\pi_{p(n)}$ is NP-complete.

With this definition, a strongly NP-complete problem cannot be solved in pseudo-polynomial time if the class P is different from the class NP. Thus, the traveling salesman problem is strongly NP-complete because the Hamiltonian cycle can polynomially transform into the traveling salesman with a distance matrix containing only 0s or 1s. Since the Hamiltonian cycle is NP-complete, traveling salesman instances involving only small numbers are also NP-complete.

Conversely, the problems that can be solved with dynamic programming, like the knapsack or the partition problem, are not strongly NP-complete. Indeed, if the sum of the weights of the n elements of a partition problem is bounded by a polynomial $p(n)$, the algorithm presented above has complexity in $O(n \cdot p(n))$ which is polynomial.

1.2.4 Other Complexity Classes

Countless other complexity classes have been proposed. Among those which are most frequently encountered in the literature and which can be described intuitively, we can cite:

NP-Hard The problems considered above are decision problems, not optimization ones. With a dichotomy algorithm, we can easily solve the optimization problem associated with a decision problem. A problem is *NP-hard* if any problem of NP can transform into this problem in polynomial time. Unlike the NP-complete class, we do not force the latter to be part of NP. Thus, an optimization problem whose decisional version is NP-complete falls into the category of NP-hard problems.

P-SPACE The problems that can be solved with a machine whose memory is limited by a polynomial in the data size belong to the class *P-SPACE*. No limit is imposed here on the computational time, which can be exponential. Thus, all the problems of NP are in P-SPACE because we can design exhaustive enumeration algorithms that do not require too much memory. An example of a problem in this class is to determine whether a two-player deterministic game is unfair, i.e., if player B is sure to lose if player A does not make mistakes. This problem is unlikely to be part of the class NP, because it is hard to imagine that a concise certificate can be given for solutions to problems of this class.

Class L The problems which can be solved with a machine whose *working memory* is bounded by a polynomial in the size of the data—by disregarding the space necessary for the storage of the problem data—are part of the *class* L. This class includes problems of finding elements in databases whose size does not fit in the computer RAM.

Class NC The class NC contains the problems that can be solved in poly-logarithmic time on a machine including a polynomial number of processors. The problems of this class can therefore be solved in parallel in a shorter time than that which is needed to sequentially read the data. The sorting of the elements of an array falls under the NC class.

Few results have been established regarding the relationships between these various complexity classes. With the exception of the obvious inclusions in the broad sense $L \subseteq P \subseteq NP \subseteq$ NP-complete $\subseteq$ P-SPACE and $NC \subseteq P$, the only strict inclusion established is $L \neq$ P-SPACE. It is conjectured that $P \neq NP$. This is a millennium problem. A deeper presentation of this topic can be found in [2].

Problems

1.1 Draw Five Segments

Try to draw five segments of lines on the plane so that each segment cuts exactly three others. Formalize this problem in terms of graphs.

1.2 O Simplification

Simplify the following expressions:

- $O(n^5 + 2^n)$
- $O(5^n + 2^{2^n})$

- $\Omega(n^2 \cdot n! + (n+2)!)$
- $\Omega(n \log(\log(n)) + 23n)$
- $O(n^{\log(n)} + n^{5+cos(n)})$
- $O(n \log(n) + n^{3-2 \cdot sin(n)})$

1.3 Turing Machine Program

Write a deterministic Turing machine program that recognizes if the substring *ane* is written on the tape. The input alphabet is $\Sigma = \{a, c, e, n\}$. Specify the tape alphabet Γ, the state set Q and the transition function δ.

1.4 Clique is NP-Complete

Show that finding a clique of a given size in a graph is NP-complete.

1.5 Asymmetric TSP to Symmetric TSP

Show that the asymmetric traveling salesman problem—the distance from city i to j can be different from the distance from city j to i—can be polynomially transformed into the symmetric TSP by doubling the number of cities.

References

1. Cook, S.A.: The complexity of theorem-proving procedures. In: Proceedings of the Third Annual ACM Symposium on Theory of Computing, pp. 151–158. ACM, New York (1971). https://doi.org/10.1145/800157.805047
2. Garey, M.R., Johnson, D.S.: Computers and Intractability — A Guide to the Theory of NP-Completeness. W. H. Freeman, New York (1979)

Chapter 2
A Short List of Combinatorial Optimization Problems

After reviewing the main definitions of graph theory and complexity theory, this chapter reviews several combinatorial optimization problems. Some of these are easy, but adding a seemingly trivial constraint can make them difficult. We also briefly review the operating principle of simple algorithms for solving some of these problems. Indeed, some of these algorithms, producing a globally optimal solution for easy problems, have strongly inspired heuristic methods for intractable ones; in this case, they obviously do not guarantee that an optimal solution is obtained.

2.1 Optimal Trees

Finding a connected sub-graph of optimal weight is a fundamental problem in graph theory. Many applications require discovering such a structure as a preliminary step. A typical example is the search for a minimum cost connected network (water pipes, electrical cables). Algorithmic solutions to this type of problem were already proposed in the 1930s [1, 2].

2.1.1 Minimum Spanning Tree

The *minimum spanning tree* problem can be formulated as follows: given an undirected network $R = (V, E, w)$ on a set V of vertices, a set E of edges with a weight function $w \to \mathbb{R}$, we are looking for a connected, cycle-free subset whose total edge weight is as small as possible. Mathematically, the minimum spanning

É. D. Taillard, *Design of Heuristic Algorithms for Hard Optimization*, Graduate Texts in Operations Research, https://doi.org/10.1007/978-3-031-13714-3_2

tree problem is not so simple to formulate. An integer linear program containing an exponential number of constraints is:

$$\text{Minimize} \quad z = \sum_{e \in E} w(e) x_e \tag{2.1}$$

$$x_e \in \{0, 1\} \qquad \forall e \in E \tag{2.2}$$

$$\sum_{e \in E} x_e = |V| - 1 \tag{2.3}$$

$$\sum_{e \in E(S)} x_e \leqslant |S| - 1 \qquad \forall S \subseteq V, S \neq \varnothing \tag{2.4}$$

where $E(S)$ is the subset of edges with both ends in the vertex subset S.

The variables x_e are constrained to be binary by (2.2). They indicate if edge e is part of the tree ($x_e = 1$) or not ($x_e = 0$). Constraint (2.3) ensures that enough edges are selected for ensuring connectivity. Constraints (2.4) eliminate the cycles in the solution. Such a mathematical model cannot be used as is, since the number of constraints is far too large. It can be used interactively. The problem is solved without cycle elimination constraints. If the solution contains a cycle on the vertices of a subset S, the constraint that eliminates it is specifically added before restarting.

Such an approach is fastidious. Fortunately, there are very simple methods for finding a minimum spanning tree. The most famous algorithms to solve this problem are those of Kruskal and Prim. They are both based on a *greedy method*. Greedy algorithms are discussed in Sect. 4.3. They build a solution incrementally from scratch. At each step, an element is included in the structure in construction, never changing the choice of this element later.

The Kruskal Algorithm 2.2 starts with a graph $T = (V, E_T = \varnothing)$. It successively adds an edge of weight as low as possible to E_T while ensuring no cycle is created.

Algorithm 2.1: (Kruskal) Building a minimum spanning tree. Efficient implementations use a special data structure for managing disjoint datasets. This is required to test if the tentative edge to add is part of the same connected component or not. In this case, the complexity of the algorithm is $O(|E| \, log \, |E|)$

Data: Undirected connected network $R = (V, E, w)$
Result: Minimum spanning tree $T = (V, E_T)$

Sort and renumber the edges by nondecreasing weight $w(e_1) \leqslant w(e_2) \leqslant \cdots \leqslant w(e_{|E|})$
$E_T = \varnothing$
for $k = 1 \ldots |E|$ **do**
  if $E_T \cup \{e_k\}$ *has no cycle* **then**
    $E_T \leftarrow E_T \cup \{e_k\}$

Algorithm 2.2: (Jarník) Building a minimum spanning tree. The algorithm was later rediscovered by Prim and by Dijkstra. It is commonly referred to as Prim or Prim-Dijkstra algorithm. For an efficient implementation, an adequate data structure must be used to extract the vertex of L with the smallest weight (Line 8) and to change the weights (Line 14). A Fibonacci heap or a Brodal queue allow an implementation of the algorithm in $O(|E| + |V| \log |V|)$

Data: Undirected connected network $R = (V, E, w)$, a given vertex $s \in V$
Result: Minimum spanning tree $T = (V, E_T)$

1 **forall** *Vertex* $i \in V$ **do**
2     $\lambda_i \leftarrow \infty$ // Cost for introducing i into T
3     $pred_i \leftarrow \varnothing$ // Predecessor of i
4 $\lambda_s = 0$; $E_T \leftarrow \varnothing$
5 $L \leftarrow V$ // List of vertices to introduce in T
6 **while** $L \neq \varnothing$ **do**
8   Remove the vertex i with the smallest λ_i from L
9   **if** $i \neq s$ **then**
10         $E_T \leftarrow E_T \cup \{pred_i, i\}$
11   **forall** *Vertex* j *adjacent to* i **do**
12   **if** $j \in L$ *and* $\lambda_j > w(\{i, j\})$ **then**
14             $\lambda_j \leftarrow w(\{i, j\})$
15             $pred_j \leftarrow i$

The Prim Algorithm 2.2 starts with a graph $T = (V' = \{s\}, E_T = \varnothing)$ and successively adds a vertex v to V' and an edge e to E_T, such that the weight of e is as low as possible and one of its ends is part of V' and the other not. Put differently, Kruskal starts with a forest with as many trees as there are vertices and seeks to merge all these trees into a single one while Prim starts with a tree consisting of a single vertex and seeks to make it growing until comprising all vertices.

2.1.2 *Steiner Tree*

The *Steiner tree* problem is very close to that of the minimum spanning tree. The sole difference is that the vertices of a subset $S \subset V$ must not necessarily appear in the tree. S is the set of *Steiner* vertices. The other ones that must belong to the tree are designated as *terminal* vertices. The Euclidean version of the Steiner tree is to connect a given set of terminal points on the plane by lines whose length is as short as possible. Figure 2.1 shows the minimum spanning tree, using solely the edges directly connecting the terminals and a Steiner tree. The weight of the minimum spanning tree may be larger than that of a Steiner tree where appropriately Steiner

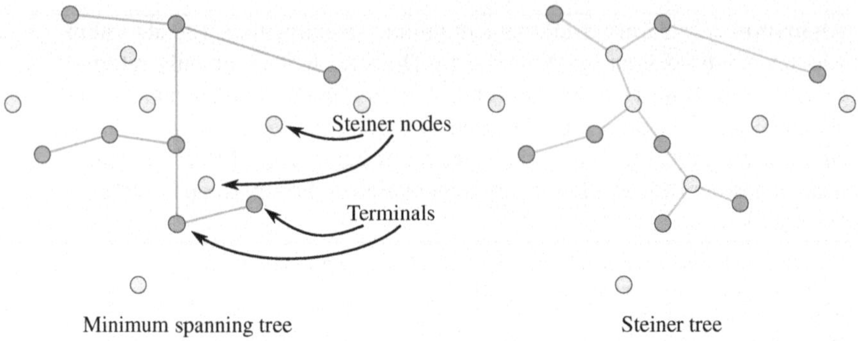

Fig. 2.1 Minimum spanning tree using only terminal nodes, which are, therefore, directly connected to each other and minimum weight Steiner tree, where additional nodes can be used

nodes are added. The combinatorial choice of vertices to add makes the problem NP-hard.

2.2 Optimal Paths

Searching for optimal paths is as old as the world. Everyone is aware of this problem, especially since cars are built with a navigation system. Knowing the current position on a transport network, the aim is to identify the best route to a given destination. The usual criterion for the optimality of the path is time, but it can also be distance, especially if it is a walking route.

2.2.1 Shortest Path

Formally, let $R = (V, E, w)$ be a directed network. We want to find a shortest walk starting at node s and ending at node t. Naturally, "shortest" is an abuse of language and designates the sum of the edge weight. The lasts can represent something other than a distance, such as a time, energy consumption, etc. Considering the algorithmic complexity, it is not more expensive to find the optimum walks from a particular node s to, or from all the vertices of V.

This formulation can be problematic in the case of a general weighting function. Indeed, if there are negative weights, the shortest walk may not exist if one has a negative length circuit. Dijkstra's algorithm is the most effective one to discover the shortest path in a network where the weighting function is not negative: $w(e) \geqslant 0 \; \forall e \in E$. It is formalized by Algorithm 2.3.

Algorithm 2.3: (Dijkstra) Searching for a shortest path from s to all other nodes in a non-negative weighting network. The red color highlights the two differences between this algorithm and Prim's one (Algorithm 2.2)

Data: Directed Network $R = (V, E, w)$ with $w(e) \geqslant 0\ \forall e \in E$, given by successor lists $succ(i)$ for each vertex $i \in V$, a given vertex s

Result: Immediate predecessor $pred_j$ of j on a shortest path from s to j, $\forall j \in V$ and length λ_j of the shortest path from s to j

1 **forall** *Vertex* $i \in V$ **do**
2     $\lambda_i \leftarrow \infty$
3     $pred_i \leftarrow \varnothing$
4 $\lambda_s = 0$
5 $L \leftarrow V$ // Vertices for which the shortest path is not definitive
6 **repeat**
8   Remove vertex i with smallest λ_i value from L
9   **forall** *Vertices* $j \in succ(i)$ **do**
10   **if** $j \in L$ *and* $\lambda_j > \lambda_i + w(i,j))$ **then**
12             $\lambda_j \leftarrow \lambda_i + w(i,j))$
13             $pred_j \leftarrow i$
14 **until** $L \neq \varnothing$

The idea behind this algorithm is to store, in a set L, the vertices for which the shortest path from the starting vertex s has not yet been definitively identified. A value λ_i is associated with each vertex i. This value represents the length of an already discovered path from s. Since we suppose non-negative weights, the node $i \in L$ with the smallest value is a new vertex for which the shortest path is definitively known. The node i can, therefore, be removed from L while checking whether its adjacent vertices could be reached with a shorter path passing through i.

For an efficient implementation, an adequate data structure must be used to extract the vertex of L with the smallest value (Line 8) and to change the values (Line 12) of its adjacent nodes. Similarly to Prim's Algorithm 2.3, a Fibonacci heap or a Brodal queue allows an implementation of the algorithm in $O(|E| + |V| \log |V|)$.

It is interesting to highlight the significant similarity between this algorithm and that of Prim 2.2 for finding a minimum spanning tree. The recipe that worked for this problem still works, with some restrictions, for discovering a shortest path. The general framework of the greedy methods, on which this recipe is based, is presented in Sect. 4.3 of the chapter devoted to constructive methods. Code 2.1 provides an implementation of Dijkstra's algorithm, in case the network is dense enough for reasonably specifying it with a square matrix.

Code 2.1 dijkstra.py Implementation of Dijkstra's algorithm for a complete network specified by a matrix (d_{ij}) providing the weight of each arc (i, j). In this case, managing L with a simple array is optimal

```
######### Dijkstra algorithm for finding all shortest paths from start
def dijkstra(n,                                         # number of cities
             d,                       # distance matrix (with no negative values)
             start):                                              # starting city

    order = [i for i in range(n)] # Cities ordered by increasing shortest path
    pred = [start] * n    # Immediate predecessor on a shortest path from start
    length = [float('inf')] * n                          # Shortest path lengths
    length[start] = 0     # Only shortest path to order[0]=start already known
    order[0], order[start] = order[start], order[0]

    for i in range(0, n - 1): # Update shortest path for neighbors of order[i]
        for j in range(i+1, n):                  # For all neighbors to update
            if length[order[i]] + d[order[i]][order[j]] < length[order[j]]:
                length[order[j]] = length[order[i]] + d[order[i]][order[j]]
                pred[order[j]] = order[i]
            # Update order if a better i+1th shortest path is identified
            if length[order[i+1]] > length[order[j]]:
              order[i+1], order[j] = order[j], order[i+1]

    return length, pred
```

Also note that Code 4.3 implements one of the most popular greedy heuristics for the traveling salesman problem. It displays exactly the same structure as Code 2.1.

When the weights can be negative, the shortest walk exists only if there is no negative length circuit in the network. Written differently, this walk must be a simple path. A more general algorithm to find shortest paths was proposed by Bellman and Ford (see Algorithm 2.4). It is based on verifying, for each arc, that the *Bellman conditions* are satisfied: $\lambda_j \leqslant \lambda_i + w(i, j)$. In other words, the length of the path from s to j should not exceed that of s to i plus the length of the arc (i, j). If it were the case, there would be an even shorter path up to j, passing through i.

The working principle of this algorithm is completely different from the greedy algorithms we have seen so far. Rather than definitively including an element to a partial solution at each step, the idea is to try to improve a complete starting solution. The last can be a very bad one, easy to build. The general framework of this algorithm is that of a local improvement method. At each step of the algorithm, the Bellman conditions are checked for all the arcs. If they are satisfied, all the shortest paths have been found. If one finds a vertex j for which $\lambda_j > \lambda_i + w(i, j)$, the best path known to the node j is updated by storing the node i as its predecessor. Making such a modification can invalidate the Bellman conditions for other arcs. It is, therefore, necessary to check again, for all arcs, if a modification has no domino effect.

A question arises: without further precaution, does an algorithm based on this labeling update stop for any entry? The answer is no: if the network has a negative length circuit, there are endless modifications. In case the network does not have a negative length circuit, the algorithm stops after a maximum of $|V|$ scans of the Bellman conditions for all the arcs of E. Indeed, if a shortest path exists, its number

Algorithm 2.4: (Bellman–Ford) Finding shortest paths from s to all other nodes in any network. The algorithm indicates if the network has a negative length circuit accessible from s, which means that the (negative) length of the shortest walk is unbounded. This algorithm is excessively simple to implement (the code is hardly longer than the pseudo-code provided here). Its complexity is in $O(|E||V|)$

Data: Directed network $R = (V, E, w)$ given with an arc list, a starting node s
Result: Immediate predecessor $pred_j$ of j on a shortest path from s to j with its length λ_j, $\forall j \in V$, or: warning message of the existence of a negative length circuit

forall $i \in V$ **do**
    $\lambda_i \leftarrow \infty$; $pred_i \leftarrow \varnothing$
$\lambda_s \leftarrow 0$
$k \leftarrow 0$ // Step counter
$Continue \leftarrow true$ // At least one λ modified at last step
while $k < |V|$ *and Continue* **do**
    $Continue \leftarrow false$
    $k \leftarrow k + 1$
    forall $arc\ (i, j) \in E$ **do**
        if $\lambda_j > \lambda_i + w(i, j)$ **then**
            $\lambda_j \leftarrow \lambda_i + w(i, j)$
            $pred_j \leftarrow i$
            $Continue \leftarrow true$
if $k = |V|$ **then**
    Warning: there is a negative length circuit that can be reached from s

of arcs is at most $|V| - 1$. Each scan of the arcs of E definitively fixes a value satisfying the Bellman condition for at least one vertex.

The Bellman–Ford algorithm is based on an improvement method with a well-defined stopping criterion: if there are still values updated after $|V|$ steps, then the network has a negative length circuit and the algorithm stops. If a scan finds out that the Bellman conditions are satisfied for all the arcs, then all the shortest paths are identified and the algorithm stops.

Seeking optimal paths appears in many applications, especially in project planning and scheduling. The problems that can be solved by dynamic programming can be formulated as finding an optimal path in a layered network. This technique uses the special network topology to find the solution without having to explicitly construct the network.

2.2.1.1 Linear Programming Formulation of the Shortest Path

It is relatively easy to formulate the problem of finding the shortest path from a vertex s to a vertex t in a network under the form of a linear program. For this purpose, a variable x_{ij} is introduced for each arc (i, j) to indicate whether the last is part of the shortest path. The formulation below may seem incomplete: indeed, the variables x_{ij} should either take the value 0 (indicating the arc (i, j) is not part of the shortest path) or the value 1 (the arc is part of it). Constraints (2.8) are sufficient: if a variable receives a fractional value in the optimal solution, it means there are several shortest paths from s to t. Constraint (2.7) imposes that there is a unit "quantity" arriving in t. This amount can be split inside the network, but each fraction must use a shortest path. Constraints (2.6) impose that the quantity arriving at any intermediate node j must depart from it. It is not required to explicitly impose that a unit quantity leaves s. Such a constraint would be redundant with (2.7). The objective (2.5) is to minimize the cost of the arcs retained.

$$\text{Minimize} \quad z = \sum_{i,j} w(i, j)x_{ij} \tag{2.5}$$

$$\sum_{i=1}^{n} x_{ij} - \sum_{k=1}^{n} x_{jk} = 0 \qquad \forall j \neq s, j \neq st \tag{2.6}$$

$$\sum_{i=1}^{n} x_{it} - \sum_{k=1}^{n} x_{tk} = 1 \tag{2.7}$$

$$x_{ij} \geqslant 0 \qquad \forall i, j \tag{2.8}$$

Another formulation of this problem is to directly look for the lengths λ_i of the shortest paths by imposing the Bellman conditions. This leads to the following linear program, which is the *dual* of the previous one.

$$\text{Maximize} \quad \lambda_t \tag{2.9}$$

$$\text{Subject} \quad \lambda_j - \lambda_i \leqslant w(i, j) \qquad \forall i, j \tag{2.10}$$

$$\text{to} \quad \lambda_s = 0 \tag{2.11}$$

Duality carries out a significant role in linear programming. Indeed, it is shown that any feasible solution to the primal problem has a value that cannot be lower than a feasible solution value to the dual. If a feasible solution value to the primal problem exactly reaches a feasible solution value to the dual, then both solutions are optimal. For the shortest path problem, the optimal λ_t value corresponds to the sum of the lengths of the arcs that must be used in an optimum path from s to t.

2.2.2 *Elementary Shortest Path: Traveling Salesman*

The shortest walk problem is poorly defined, because of the negative length circuits. However, one could add a very natural constraint, which makes it perfectly defined:

look for the shortest *elementary* path from a particular node s to all the others. It is recalled that an elementary path visits each vertex at most once. In this case, even if there are negative length circuits, the problem has a finite solution. Unfortunately, adding this little constraint makes the problem difficult. Indeed, it can be shown that the traveling salesman problem, notoriously NP-hard, can transform polynomially into the elementary shortest path problem.

The traveling salesman problem (TSP) is the archetype of hard combinatorial optimization, on the one hand, because of the simplicity of its formulation and, on the other hand, because it appears in many applications, particularly in vehicle routing.

The first practical application of the traveling salesman problem is clearly finding a shortest tour for a trading clerk. In the nineteenth century, Voigt edited a book exhibiting how to make a round trip in Germany and Switzerland [5].

There are many practical applications to this problem. For instance, Sect. 2.2.3, shows that vehicle routing implies solving many traveling salesman instances. As presented further (see Sect. 3.3.1), it can also appear in problems that have nothing to do with routing.

In combinatorial optimization, the TSP is most likely the one that has received the most attention. Large Euclidean instances—more than 10,000 nodes—have been optimally solved. There are solutions that do not deviate from more than a fraction of a percent from the optimum for instances with several million cities. Since this problem is NP-hard, there are much smaller examples that cannot be solved by exact solution methods. The TSP polynomially transforms into the shortest elementary path as follows.

A vertex is duplicated in two vertices s and t and the weight $w(i, j)$ of all the arcs is replaced by $w(i, j) - M$, where M is a positive constant larger than the largest weight of an arc. If there is no arc between s and t, the shortest elementary path from s and t corresponds to a minimum tour length for the traveling salesman. Figure 2.2 illustrates the principle of this transformation. Knowing that the TSP is NP-hard, it proves that the shortest elementary path is NP-hard too.

2.2.2.1 Integer Linear Programs for the TSP

There are numerous integer linear programs modeling the TSP. Two of the best known are presented here.

Dantzig-Fulkerson-Johnson

The Dantzig-Fulkerson-Johnson formulation introduces an exponential number of sub-tour elimination constraints. The binary variables x_{ij} take the value 1 if the arc

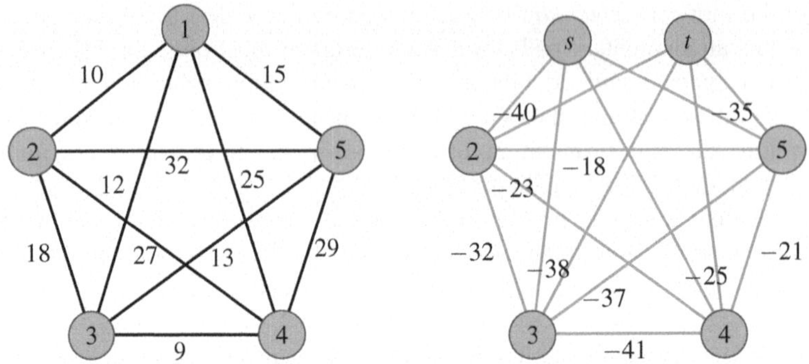

Fig. 2.2 Polynomial transformation from a traveling salesman into an elementary shortest path. Vertex 1 is duplicated and the weight of each edge is set to the original weight minus 50. Finding the shortest elementary path from s to t is equivalent to finding the optimal TSP tour in the original network

(i, j) is used in the tour and 0 otherwise.

$$\text{Minimize} \qquad z = \sum_{(i,j)} w(i, j)x_{ij} \tag{2.12}$$

$$x_{ij} \in 0, 1 \quad \forall i, j \tag{2.13}$$

$$\sum_{i=1}^{n} x_{ij} = 1 \qquad \forall j \tag{2.14}$$

$$\sum_{j=1}^{n} x_{ij} = 1 \qquad \forall i \tag{2.15}$$

$$\sum_{(i,j)\in E(S)} x_{ij} \leqslant |S| - 1 \quad \forall S \subsetneq V, S \neq \varnothing \tag{2.16}$$

Constraints (2.14) impose to enter exactly once in each city. Constraints (2.15) impose to come out exactly once from each city. Constraints (2.16) ensures that no proper subset S contains a sub-tour.

Compared to the linear program for finding a minimum weight tree, it differs only in Constraints (2.14) and (2.15) which replace Constraint (2.3).

Miller–Tucker–Zemlin

The Miller–Tucker–Zemlin formulation replaces the exponential number of constraints (2.16) by a polynomial number of constraints and introducing $|V| - 1$ continuous variables u_i, $(i = 2 \ldots |V|)$. The new variables provide tour ordering. If

$u_i < u_j$, then city i is visited before city j. In this formulation, constraints (2.13)–(2.15) are retained and constraints (2.16) are replaced by:

$$u_i - u_j + |V|x_{ij} \leqslant |V| - 1 \qquad 2 \leqslant i \neq j \leqslant |V| \tag{2.17}$$

$$1 \leqslant u_i \leqslant |V| - 1 \qquad 2 \leqslant i \leqslant |V| \tag{2.18}$$

This integer linear program is probably not the most efficient one, but it has relatively few variables and constraints.

2.2.3 *Vehicle Routing*

Problems using the traveling salesman as a sub-problem naturally appear in the *vehicle routing* problem (VRP). In its simplest form, the last can be formulated as follows: let V be a set of customers requesting quantities q_i of goods ($i = 1, \ldots, |V|$). They are delivered by a vehicle with capacity Q, starting from and returning to a warehouse d. The customers must be split into m subsets $V_1, \ldots V_m$ such that $\sum_{i \in V_i} q_i \leqslant Q$. For each subset $V_j \cup \{d\}, (j = 1, \ldots, m)$, a traveling salesman tour as short as possible must be determined. Figure 2.3 illustrates a solution of a small VRP instance.

This problem naturally occurs for delivering or collecting goods and in home service planning. In concrete applications, many complications exist:

- The number m of tours can be fixed or minimized;
- The maximum length of the tours can be limited;
- The clients specify one or more time windows during which they should be serviced;
- The goods can be split implying multiple passages at the same client;
- A tour can both collect and deliver goods;
- There is more than one warehouse;
- Warehouses are hosting heterogeneous fleets of vehicles;
- The warehouses locations can be chosen;
- etc.

Since the problem is to find the service order of customers, the problem is also referred to as "Vehicle Scheduling."

2.3 Scheduling

Scheduling is to determine the order to process a number of operations. Their processing consumes resources, for instance, time on a machine. Operations that need to be processed in a specific order are grouped into jobs. The purpose of

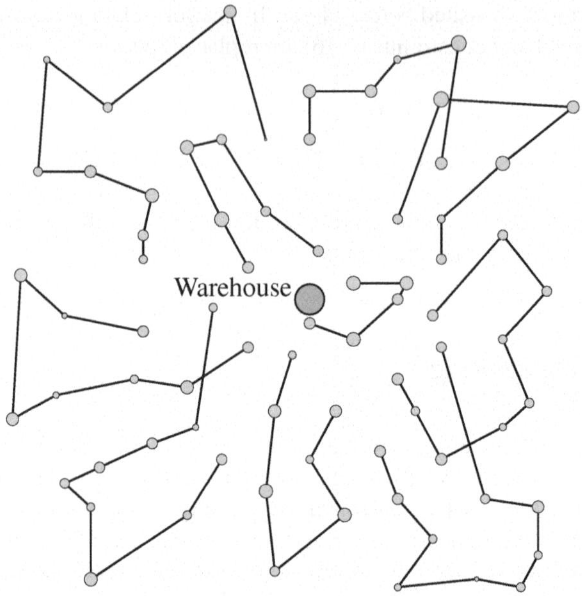

Fig. 2.3 Vehicle routing problem instance. Trips from and to the warehouse are not drawn for not overloading the illustration. This solution was discovered by means of a taboo search, but it took decades before its optimality was proven. This gives an idea of the difficulty of the problem

scheduling is to optimize resource consumption. Various optimization criteria are commonly used: minimizing the makespan; minimizing the total time; minimizing the average delay; etc. A frequent constraint in scheduling is that a resource cannot perform several operations simultaneously and that two operations of a job cannot be performed simultaneously. Operations may include various features according to applications:

Resource An operation must take place on a given resource or subset of resources or must require several resources simultaneously.

Duration Processing an operation takes time, which may depend on the operating resource.

Set-up time Before performing an operation, the resource requires a set-up time depending on the previously completed operation.

Interrupt After an operation has started, it can be suspended before ending.

Pre-emption A resource can interrupt an operation to process another one.

Waiting time There can be either a waiting time between two successive operations of the same task or a waiting time is prohibited.

Release date An operation cannot take place before being available.

Deadline An operation cannot be processed after a given date.

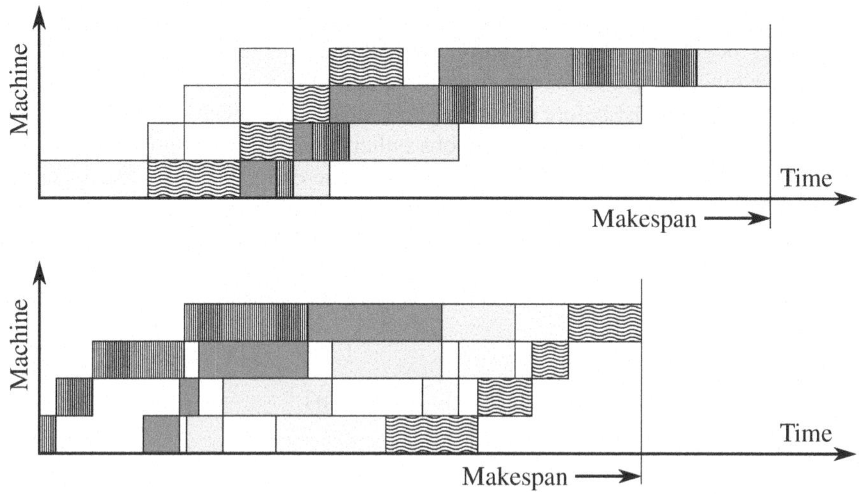

Fig. 2.4 Permutation flowshop scheduling. Gantt chart for a small instance with 4 resources and 5 jobs. Top: non-optimal schedule with the earliest starting time for each operation. Bottom: optimal scheduling with the latest starting time

In addition, resources may have a variety of features. They can be mobile in the case of carriers, resulting in colliding problems. There may be several machines of the same type, machines that can perform different operations, etc.

2.3.1 Permutation Flowshop Scheduling

A fundamental scheduling problem is the *permutation flowshop*. This problem occurs, for example, in an assembly line in which the n jobs must be successively processed on the machines $1, 2, \ldots, m$, in that order. A job j must, therefore, undergo m operations which take a time t_{ij}, $(i = 1, \ldots, m, j = 1, \ldots, n)$. The goal is to find the order to process the job in the assembly line. Written differently, to find a permutation of the job such that the last job on the last machine finishes as early as possible. There is a buffer that may store jobs between each machine. Hence, the jobs can possibly wait for the next machine to finish processing a job that has arrived earlier. A convenient way to represent a scheduling solution is the *Gannt chart*. The x-axis represents time and the y-axis represents resources.

Figure 2.4 provides both Gannt charts of a non-optimal solution, where each operation is planned as early as possible as well as an optimal scheduling where each operation starts as late as possible.

For problem instances with only 2 machines, there is a greedy algorithm finding an optimal solution to this problem. The operations are ordered by increasing durations and put in a list. The operation with the shortest duration is first selected.

If this operation takes place on the first machine, the corresponding job is placed at the very beginning of the sequence. Else, if the operation takes place on the second machine, the job is placed at the very end of the sequence. The operation is removed from the list before examining the subsequent operation. The sequence is thus completed by dispatching the jobs either after the block processed at the beginning of the sequence or before the block at the end. As soon as the instance has more than 2 machines, the problem is NP-hard. A mixed integer linear program for the permutation flowshop is as follows:

$$\text{Minimize } d_\omega \tag{2.19}$$

$$d_{mj} + t_{mj} \leqslant d_\omega \quad (j = 1 \ldots n) \tag{2.20}$$

$$d_{ij} + t_{ij} \leqslant d_{i+1j} \quad (i = 1, \ldots m-1, j = 1 \ldots n) \tag{2.21}$$

$$d_{ij} + t_{ij} \leqslant d_{ik} + M \cdot (1 - y_{jk}) \quad (i = 1, \ldots m, j = 1 \ldots n, j < k = 2 \ldots n) \tag{2.22}$$

$$d_{ik} + t_{ik} \leqslant d_{ij} + M \cdot y_{jk} \quad (i = 1, \ldots m, j = 1 \ldots n, j < k = 2 \ldots n) \tag{2.23}$$

$$d_{ij} \geqslant 0 \quad (i = 1, \ldots m, j = 1 \ldots n) \tag{2.24}$$

$$y_{jk} \in \{0, 1\} \quad (j = 1 \ldots n, j < k = 2 \ldots n) \tag{2.25}$$

Objective (2.19) is to minimize the makespan d_ω. The variable d_{ij} corresponds to the starting time of job j on machine i. Constraints (2.20) require that the end of the process of each object j on the last machine occurs not later than the makespan. A job j must have finished its processing on a machine i before being processed by the machine $i + 1$ (2.21). The y_{jk} variables indicate whether the job j should be processed before the job k. Only $n \cdot (n - 1)/2$ of these $y_{..}$ variables are introduced, since y_{kj} should take the complementary value $1 - y_{jk}$. Both Constraints (2.22) and (2.23) involve a large constant M for expressing *disjunctive constraints*: either the job j is processed before the job k or k before j. If $y_{jk} = 1$, j is processed before k and Constraints (2.23) are trivially satisfied for any machine i, provided M is large enough. Conversely, if $y_{jk} = 0$, Constraints (2.22) are trivially satisfied while Constraints (2.23) require finishing the processing of k on the machine i before the latter can start the processing of j.

2.3.2 *Jobshop Scheduling*

The *jobshop scheduling* problem is somewhat more general. Each job undergoes a certain number of operations, each of them being processed by a given machine. The operation sequence for a job is fixed, but different jobs do not necessarily have the same sequence and the jobs are not required to be processed by all machines.

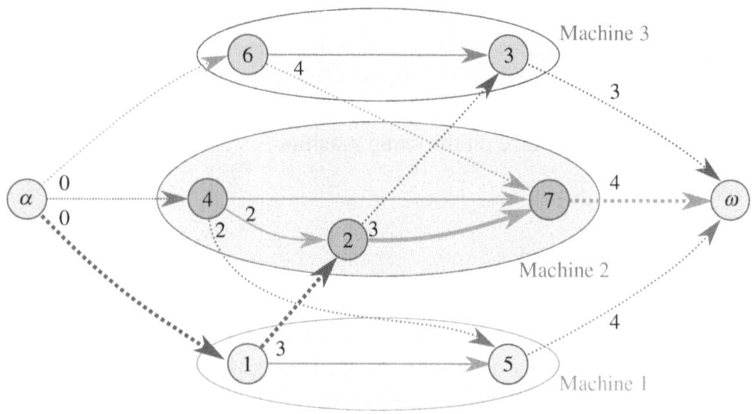

Fig. 2.5 Graph corresponding to the solution of a jobshop instance with three machines. One job undergoes 3 operations while two others have only 2. The weighting of the arcs corresponds to the duration of the corresponding operation. The arcs representing the precedence relations of the operations belonging to the same job are in dotted lines. The longest path from α to ω is shown in bold. It is referred to as the critical path

Figure 2.5 illustrates how to express this problem in terms of a graph: Each operation is associated with a vertex. Two fictitious vertices-operations are added: *start* (α) and *end* (ω). If operation k immediately follows operation i on the same job, an arc (i, j) is introduced. The length of the arc is t_i, corresponding to the duration of operation i. Arcs of length 0 are added from *start* to the first operations of each job. Arcs with a length corresponding to the duration of the last operation of each job are connecting the *end* vertex.

All operations taking place on the same machine are forming a clique. The goal of the problem is to direct the edges of these cliques to minimize the length of the longest path from *start* to *end*.

An integer linear program for the jobshop is as follows:

$$\text{Minimize } d_\omega \tag{2.26}$$

$$d_i + t_i \leqslant d_j \quad \forall (i, j) \tag{2.27}$$

$$d_i + t_i \leqslant d_\omega \quad \forall i \tag{2.28}$$

$$d_i + t_i \leqslant d_k + M \cdot (1 - y_{ik}) \quad \forall i, k \text{ on the same machine} \tag{2.29}$$

$$d_k + t_k \leqslant d_i + M \cdot y_{ik} \quad \forall i, k \text{ on the same machine} \tag{2.30}$$

$$d_i \geqslant 0 \quad \forall i \tag{2.31}$$

$$y_{ik} \in \{0, 1\} \quad \forall i, k \text{ on the same machine} \tag{2.32}$$

The variable d_i is the starting time of operation i. The goal is to minimize the makespan d_ω (the starting time of the dummy operation ω). Constraints (2.27) require that operation i must be completed before starting operation j if i precedes j for a given job. Constraints (2.28) require that the end of processing times for all operations precede the end of the project. The variables y_{ik} associated with the disjunctive constraints (2.29) and (2.30) determine whether operation i precedes operation k, which takes place on the same machine.

2.4 Flows in Networks

The concept of flow arises naturally when considering material, people, or electricity that must be transported over a network. In each node one must have the equivalent to Kirchhoff's current law: the amount of flow coming to a node must be equal to the amount going out of that node.

The most elementary form of flow problems is as follows. Let $R = (V, E, w)$ be a network. Flows values x_{ij} passing through the arcs $(i, j) \in E$ are sought such that the sum of the flows issuing from a particular source-node s to reach a sink-node t is maximized. The conservation of flows must be respected: the sum of the flows entering a vertex must equal that of exiting the vertex, except for s and t. Then, the flows x_{ij} cannot be negative and cannot exceed the positive value $w(i, j)$ associated with the arcs. To solve this problem, Ford and Fulkerson proposed the relatively simple Algorithm 2.5.

Algorithm 2.5: (Ford and Fulkerson) Maximum flow from s to t

```
Input: Oriented network R = (V,E,w), a source-node s and a sink-node t
Result: Maximum flow from s to t
Starts with a null flow in all arcs
repeat
    Build the residual network R* corresponding to the current flow
    if There is a path from s to t in R* then
        Find the maximal possible flow from s to t in R* along this path
        Superimpose this flow on the current flow (diminish the flow in the arcs (i,j) of R
          appearing as (j,i) arcs on the path in R*)
until there is no path from s to t in R*
```

It is based on an improvement method: its start from a null flow (which is always feasible) increasing it at each step along a path from s to t until reaching the optimum flow. The first step of this algorithm is illustrated in Fig. 2.6. However, we can be blocked in a situation where there is no augmenting path from s to t while not having the maximal flow.

To overcome this difficulty, it should be noted that we can virtually increase the flow from a vertex j to a vertex i by decreasing it from i to j. Therefore, at each stage of the algorithm, a *residual network* is considered.

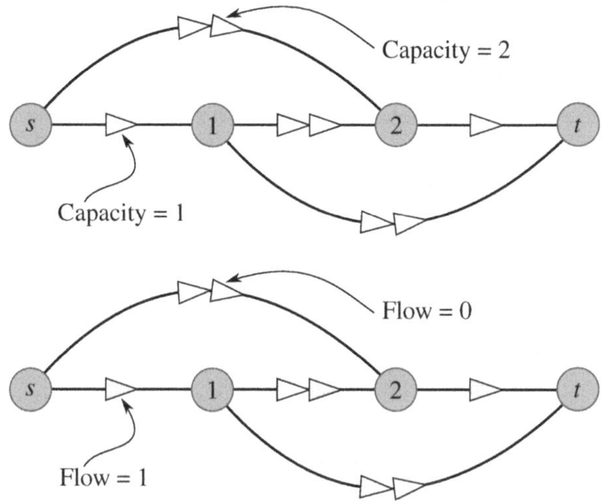

Fig. 2.6 Representation of a flow problem in terms of graphs. An empty triangle indicates an unused flow capacity. A unit flow passing through an arc is indicated by a filled triangle. The Ford and Fulkerson algorithm starts from a null flow (top) and finds the largest increase along a path from s to t. For this example, the first path discovered is $s - 1 - 2 - t$. After augmenting the flow along this path, there is no direct augmenting path (bottom)

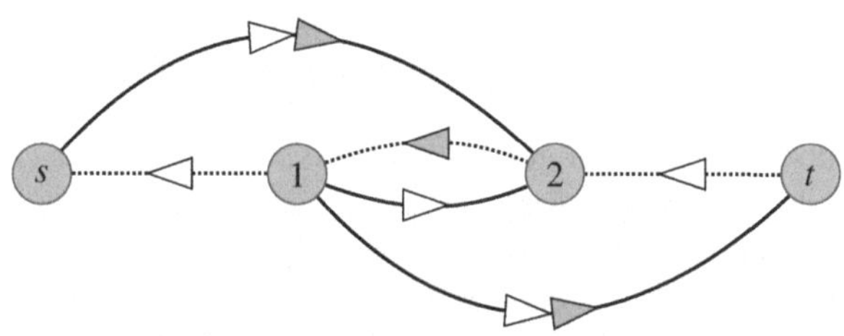

Fig. 2.7 Residual network associated with the flow of Fig. 2.6

The last is built as follows: an arc (i, j) with capacity $w(i, j)$ and with a flow x_{ij} passing through is replaced by two arcs, one from the vertex i to j with capacity $w(i, j) - x_{ij}$ (only if this value is strictly positive) and the other one from j to i with capacity x_{ij}. Figure 2.7 illustrates this principle.

Once a flow is found in the residual network, it is superimposed on the flow obtained previously. This is shown in Fig. 2.8.

The complexity of this algorithm depends on the network size. Indeed, we have to seek a path from s to t for each increasing flow. It also depends on the number of augmenting paths. Unluckily, the increase can be marginal in the worst case. For networks with integer capacities, the increase can be only 1. If the maximum capacity of an edge is m, the complexity of the algorithm is in $O(m \cdot (|E| + |V|))$. If m is small, for example, if the capacity of all the arcs is 1, the Ford and Fulkerson algorithm is fast.

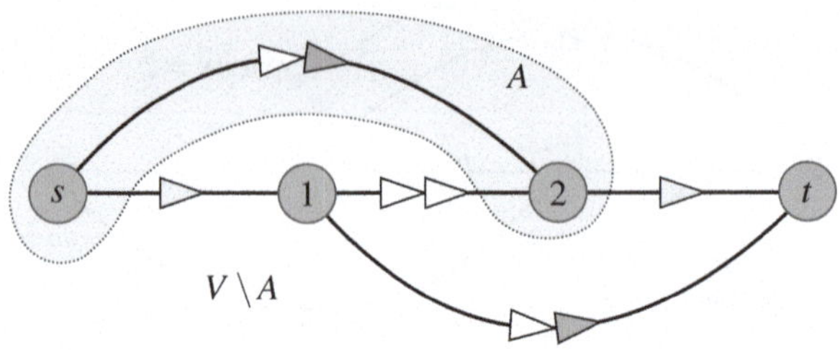

Fig. 2.8 The flow found in the residual network (Fig. 2.7) is superimposed on the previous flow (Fig. 2.6). The subset $A \subset V$ cuts s from t and the sum of the capacities of all the arcs going out of A is equal to the value of the flow. This proves the optimality of the flow

We will see in Sect. 2.8.1 how to solve the edge coloring problem of a bipartite graph by solving maximum flow problems in a network where all the capacities are 1. Its complexity can be significantly diminished by using a breadth-first search as a sub-algorithm to discover a path from s to t in the residual network. Hence, the flow is increased along the shortest path at each step. This improvement has been proposed by Edmonds and Karp.

Since the number of arcs of the path cannot decrease from one step to the next, no more than $|E|$ steps are performed with a given number of arcs. Since the number of arcs of a path is between 1 and $|V|$, we deduce that the complexity can be reduced to $O(|V||E|^2)$. In the case of a dense network (with $|E| \in O(|V^2|)$), the complexity simplifies to $O(|V|^5)$. Many algorithms have been proposed for solving the maximum flow problem. For general networks, the algorithmic complexity has been recently reduced to $O(|V||E|)$.

For many applications, each unit of flow in arc (i, j) costs $c(i, j)$. We, therefore, consider a network $R = (V, E, w, c)$, where $w(i, j)$ is the capacity of the arc (i, j) and $c(i, j)$ the cost of a unit flow through this arc. Then arises the problem of the maximum flow at minimum cost. This problem can be solved with Algorithm 2.6 of Busacker and Gowen, provided the network does not contain a negative cost circuit.

Algorithm 2.6: (Busacker and Gowen) Maximum flow from s to t with minimum cost

Input: Oriented network $R = (V, E, w, c)$ without negative circuit, a source-node s and a sink-node t

Result: Maximum flow with minimum cost from s to t

Start with a null flow in all the arcs
repeat
    Build the residual network R^* relative to the current flow
    if *A path form s to t exists in R^** **then**
        Find the maximal possible flow through the shortest path from s to t in R^*
        Superimpose this flow on the current flow
until *there is no path from s to t in R^**

As noted for the algorithms of Prim and Dijkstra, there is a very slight difference between Algorithms 2.5 and 2.6. Once more, we do not alter a winning formula! When constructing the residual network, the costs should be taken into account. If there is a flow $x_{ij} > 0$ through the arc (i, j), then the residual network includes an arc (i, j) with capacity $w(i, j) - x_{ij}$ (provided this capacity is positive) with an unchanged cost $c(i, j)$ and a reversed arc (j, i) with capacity x_{ij} and cost $-c(i, j)$.

In the general case, finding the maximum flow with minimum cost is NP-hard. Indeed, the TSP can be polynomially transformed into this problem. The transformation is similar to that of the shortest elementary path (see Fig. 2.2).

The algorithms for finding the optimal flows presented above can solve many problems directly related to flow management, like electric power distribution or transportation problems. However, they are chiefly exploited for solving assignment problems (see next Chapter for modeling the linear assignment as a flow problem).

2.5 Assignment Problems

Assignment or matching problems occur frequently in practice. This is to match the elements of two different sets like teachers to classes, symbols to keyboard keys, and tasks to employees.

2.5.1 *Linear Assignment*

The *linear assignment* problem can be formalized as follows. Given an $n \times n$ matrix of costs $C = (c_{iu})$ each element $i \in I$ must be assigned to an element $u \in U$ $(i, u = 1, \ldots, n)$ in such a way that the sum of costs (2.33) is minimized. This problem can be modeled by an integer linear program:

$$\text{Minimize} \quad \sum_{i=1}^{n} \sum_{u=1}^{n} c_{iu} x_{iu} \tag{2.33}$$

Subject to

$$\sum_{i=1}^{n} x_{iu} = 1 \quad u = 1, \ldots, n \tag{2.34}$$

$$\sum_{u=1}^{n} x_{iu} = 1 \quad i = 1, \ldots, n \tag{2.35}$$

$$x_{iu} \in \{0, 1\} \quad (i, u = 1, \ldots, n) \tag{2.36}$$

Constraints (2.34) ensure to assign exactly one element of U to each element of I. Constraints (2.35) ensure to assign exactly one element of I to each element of U. Hence, these two sets of constraints ensure a perfect matching between the elements of I and U. The integrality constraint (2.36) prevents elements of I to share fractions of elements of U.

A more concise formulation of the linear assignment problem is to find a permutation p of the n elements of the set U which minimizes $\sum_{i=1}^{n} c_{ip_i}$. The value p_i is the element of U assigned to i.

2.5.2 *Generalized Assignment*

In some cases, it is not necessary to have a perfect matching. This is particularly the case if the size of the sets I and U differ. To fix the ideas, let I be a set of n tasks to be performed by a set U of m employees, with $m < n$. If employee u performs task i, the cost is c_{iu} and the employee needs a time of w_{iu} to perform this task. Each employee u has a time budget limited by t_u.

This problem, called the *generalized assignment problem*, occurs in various practical situations. For instance, it is closely related to the distribution of the loads between vehicles for the vehicle routing problems presented in Sect. 2.2.3. The generalized assignment problem can be modeled by the integer linear program:

$$\text{Minimize} \quad \sum_{i=1}^{n} \sum_{u=1}^{m} c_{iu} x_{iu} \tag{2.37}$$

Subject to

$$\sum_{i=1}^{n} w_{iu} x_{iu} \leqslant t_u \quad u = 1, \ldots, n \tag{2.38}$$

$$\sum_{u=1}^{m} x_{iu} = 1 \quad i = 1, \ldots, n \tag{2.39}$$

$$x_{iu} \in \{0, 1\} \quad (i, u = 1, \ldots, n) \tag{2.40}$$

This small modification of the assignment problem makes it NP-hard.

2.5.3 *Knapsack*

A special case of the generalized assignment problem (see Exercise 2.9) is the *knapsack* problem. It is certainly the simplest NP-hard problem to formulate in terms of integer linear programming:

$$\text{Maximize} \quad \sum_{i=1}^{n} c_i \cdot x_i \tag{2.41}$$

Subject to:

$$\sum_{i=1}^{n} v_i \cdot x_i \leqslant V \tag{2.42}$$

$$x_i \in \{0, 1\} \quad (i = 1, \dots, n) \tag{2.43}$$

Items of volume v_i and value c_i, $(i = 1, \dots, n)$ can be put into a knapsack of volume V. The volume of the items put in the knapsack cannot be larger than V. The value of the selected items must be maximized.

This problem is used in this book to illustrate the working principles of a few methods. The reader interested in knapsack problems and extensions like bin-packing, subset-sum, and generalized assignment can refer to [4].

2.5.4 *Quadratic Assignment*

There is another assignment problem where the elements of the set I have interactions with each other. An assignment chosen for an element $i \in I$ has repercussions for the set of all the elements of I. Let us take the example of assigning n offices to a set of n employees.

In the linear assignment problem, the c_{iu} values only measure the interest for the employee i to be assigned the office u. Assigning the office u to the employee i has no other consequence than the office u is no longer available for another employee. In practice, employees are required to collaborate, which causes them to have to move from one office to another. Let a_{ij} be the frequency the employee i meets the employee j. Let b_{uv} the travel time from office u to office v. If we assign the office v to the employee j and the office u to the employee i, the last loses a time given by $a_{ij} \cdot b_{uv}$ for traveling, on average. Minimizing the total time lost can be modeled by the following quadratic 0-1 program, where the variable x_{iu} takes the value 1 if the employee i occupies the office u and the value 0 otherwise:

$$\text{Minimize} \quad \sum_{i=1}^{n} \sum_{j=1}^{n} \sum_{u=1}^{n} \sum_{v=1}^{n} a_{ij} b_{uv} x_{iu} x_{jv} \tag{2.44}$$

Subject to

$$\sum_{i=1}^{n} x_{iu} = 1 \quad u = 1, \dots, n \tag{2.45}$$

$$\sum_{u=1}^{n} x_{iu} = 1 \quad i = 1, \dots, n \tag{2.46}$$

$$x_{iu} \in \{0, 1\} \quad (i, u = 1, \dots, n) \tag{2.47}$$

This formulation brings out the quadratic side of the objective due to the product of the variables $x_{iu} \cdot x_{jv}$. Constraints (2.45)–(2.47) are typical for assignment problems. So, this problem is called the *quadratic assignment* problem. A more concise model is searching for a permutation p that minimizes

$$\sum_{i=1}^{n} \sum_{j=1}^{n} a_{ij} \cdot b_{p_i p_j}$$

Many practical applications can be formulated as a quadratic assignment problem (QAP):

Allocation of offices to employees This is the example just cited formerly.

Allocation of blocks in an FPGA A Field Programmable Gate Array requires connecting logic blocks on a silicon chip. These blocks allow implementing logic equations, multiplexers, or memory elements. Configuring an FPGA starts by establishing the way the modules must be connected. This can be described by means of a routing matrix $A = (a_{ij})$ which gives the number of connections between modules i and j. Next, each module i must be assigned a logic block p_i on the chip. Since the signal propagation delay depends on the length of the links, the assignment must be carefully performed. Therefore, knowing the length b_{uv} of the link between logic blocks u and v, the problem of minimizing the sum of the propagation times is a quadratic assignment problem.

Configuring a keypad To enter text on a cellular phone keypad, the 26 letters of the alphabet, as well as space, have been assigned to the keys $0, 2, 3, \dots, 9$. As standard, these 27 signs are distributed according to the configuration of Fig. 2.9a.

Assume that typing a key takes one unit of time, moving from one key to another takes two units of time, and finally that we have to wait 6 units of time before we can start typing a new symbol positioned on the same key. Then it takes 70 units of time to type the text “a ce soir bisous.”

Indeed, it takes 1 unit to enter the “a” on key 2, then moving to key 0 takes two units, then 1 unit to press once for space, then 2 units to move to key 2 again and 3 units for seizing “c,” etc. With the optimized keyboard (for the French language) given in Fig. 2.9, it takes only 51 units of time, almost a third less. This optimized keyboard was obtained by solving a quadratic assignment problem for which the a_{ij} coefficients represent the frequency of occurrence of the symbol j after the symbol i in a typical text and b_{uv} represents the time between the typing of a symbol placed in position u and another in position v.

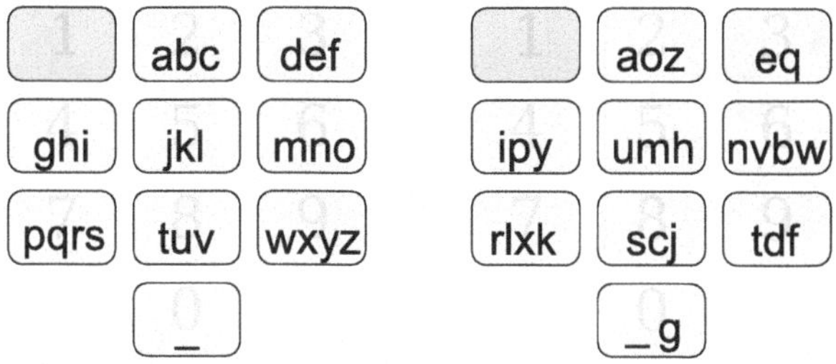

(a) (b)

Fig. 2.9 Standard cellular phone keyboard and keyboard optimized for the French language. (**a**) Standard keyboard. (**b**) Optimized keyboard

The quadratic assignment problem is NP-hard. In practice, this is one of the most difficult of this class. Yet, examples of problems of size $n = 30$ are not optimally solved. Many NP-hard problems can be transformed into quadratic assignment problems. Without being exhaustive, let us mention the traveling salesman, the linear ordering, the graph bipartition or the stable set problems. Naturally, modeling one of these problems under the form of a quadratic assignment is undoubtedly not leading to the most efficient solving method!

2.6 Stable Set

Finding the largest independent set—maximal stable set—is a classical graph theory problem. This problem is NP-difficult. Section 1.2.3.4 presents a polynomial transformation of satisfiability into stable set. The latter is equivalent to finding the largest subset of mutually adjacent nodes—a maximum clique—in the complementary graph. A variant of the maximum stable set is the maximum weight stable set, when weights are associated with vertices. In this case, we are looking for a subset of independent vertices whose sum of the weights is as high as possible. Naturally, if the weights are all the same, this variant is equivalent to the maximum stable set.

This problem appears in several practical applications: map labeling, berth allocation to ships or assigning flight level to aircrafts. This is discussed in Sect. 3.3.3.

2.7 Clustering

Like graph theory, clustering is a very useful modeling tool. There is a myriad of applications of clustering. Let us quote social network analysis, medical imaging, market segmentation, anomaly detection, and data compression. *Clustering* or

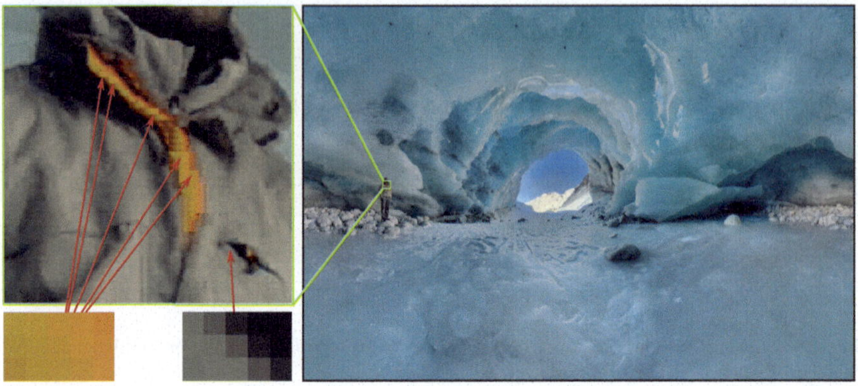

Fig. 2.10 Compression by vector quantization. An image compression technique creates clustering instances with millions of items and thousands of clusters. Here, the initial image was divided into blocks of $b = 3 \times 5$ pixels. We next looked for a palette of 2^k "colors" seen as vectors of length $b \times 3$, each pixel being characterized by its red, green, and blue brightness. For this image, we chose $k = 14$. Two of the $2^{14} = 16{,}384$ "colors" are shown. The palette was found with a clustering method, each color being a centroid. Each block of the initial image is replaced by the most similar centroid. As a block of the compressed image can be represented by k bits, k/b bits are enough to encode one pixel

unsupervised classification consists in grouping items that are similar and separating those that are not. There are specific algorithms to perform these tasks automatically. Figure 2.10 gives an example of a large clustering instance where a decomposition method, such as those presented in Sect. 6.4.2 is required. Image compression by vector quantization involves dealing with instances with millions of elements and thousands of clusters.

The supervised classification considers labeled items. It is frequently used in artificial neural networks. These techniques are outside the scope of this book, as are phylogenetic trees, popularized by Darwin in the nineteenth century.

Creating clusters supposes we can quantify the dissimilarity $d(i, j) \geqslant 0$ between two elements i and j belonging to the set E we are trying to classify. Often, the function $d(i, j)$ is a *distance*, (with symmetry: $d(i, j) = d(j, i)$, separation: $d(i, j) = 0 \iff i = j$ and triangular inequality: $d(i, k) \leqslant d(i, j) + d(j, k)$), but not necessarily. However, to guarantee the stability of the algorithms, let us suppose that $d(i, j) \geqslant 0$ and $d(i, i) = 0, \forall i, j \in E$. As soon as we have such a function, the homogeneity of a group $G \subset E$ can be measured. Several definitions have been proposed. Figure 2.11 shows some dissimilarity measures for a group of 3 elements.

Diameter of a group Maximum value of the function $d(i, j)$ for two entities i and j belonging to G: $max_{i,j \in G} d(i, j)$.

Star Sum of the dissimilarities between the most representative element of G and the others: $min_j \sum_{i \in G} d(i, j)$. When this element j must be in G, j is called a *medoid*. For instance, if the elements are characterized by two numeric values and $G = \{(0, 1), (3, 3), (5, 0)\}$, the point j of $\mathbb{R}^2$ minimizing the sum of the

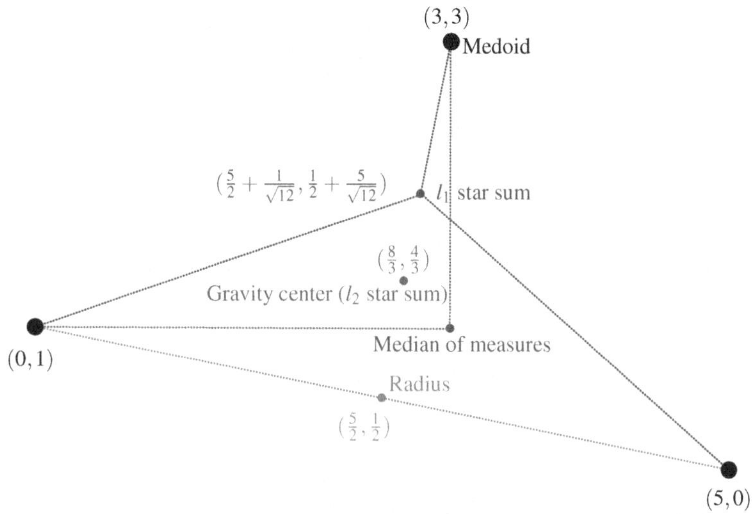

Fig. 2.11 Optimal points for various homogeneity measures of a group with 3 elements in $\mathbb{R}^2$

dissimilarities is $(5/2 + 1/\sqrt{12}, 1/2 + 5/\sqrt{12})$ with norm l_1 whereas it is the point $(8/3, 4/3)$ if we consider the standard l_2 norm. In general, there is no analytical formula to find the central point with the l_1 norm (it can be numerically estimated). For the l_2 norm, the best point is the center of gravity or centroid (mean measurements on each coordinate). The medoid of G is (3, 3).

Radius Maximum dissimilarity between the most representative element j and another of G: $min_j max_{i \in G} d(i, j)$. This element is not necessarily part of G. For instance, we can take the median (on each characteristic) or the point which minimizes any dissimilarity function. Using the numerical example above, the median of the measures is (3, 1), which does not belong to G. By taking the ordinary distance (l_1 norm) or squared distance (l_2 norm) as a dissimilarity measure, the point (5/2, 1/2) minimizes the radius of G.

Clique Sum of dissimilarities between all pairs of elements of G: $\sum_{i \in G} \sum_{j \in G} d(i, j)$.

Several definitions have been proposed to measure heterogeneity existing between two groups G and H:

Separation Minimum distance between two elements belonging to different groups: $min_{i \in G, j \in H} d(i, j)$.

Cut Sum of dissimilarities between elements of two different groups: $\sum_{i \in G} \sum_{j \in H} d(i, j)$.

Normalized cut Average of the dissimilarities between elements of two different groups: $\sum_{i \in G} \sum_{j \in H} d(i, j)/(|G| \cdot |H|)$.

Once a criterion of homogeneity or heterogeneity has been defined, we can formulate the problem of classification into p groups $G_1, \ldots G_p$ by an optimization problem using a global objective:

- Maximize the smallest separation (or the smallest cut) between elements of different groups;
- Minimize the largest diameter (or the largest radius, or the largest clique or even the largest star) of a group;
- Minimize the sum of the stars (or the sum of the diameters, radius, clique).

2.7.1 k-Medoids or p-Median

The *k-medoids* problem is one of the best known in unsupervised classification. Frequently, the terms p-median or k-medians are used instead of k-medoids in location theory and statistics. Using the definitions presented above, it is about minimizing the sum of the stars. In other words, we have to find the k elements $c_1, \ldots, c_k$ of E minimizing: $\sum_k \sum_{i \in E} min_{r=1,\ldots,k} d(i, c_r)$. This problem is NP-hard.

A well-known heuristic algorithm is the *Partition Around Medoids* (PAM 2.7). This algorithm is a local search improvement method. Various authors have proposed variations of this heuristic—while calling it PAM, which causes some confusion. The method originally proposed [3] is a local search with best improvement policy (see Sect. 5.1.2). This method requires an initial position of the centers. Different authors have suggested various methods to build an initial solution. Generally, greedy algorithms are used (see Sect. 4.3). Algorithm 2.7 does not specify how the latter is obtained; it is simply assumed that an initial solution is provided. A random solution perfectly works.

Algorithm 2.7: (PAM) Local search for clustering around medoids

Input: Set E of items with a dissimilarity function $d(i, j)$ between items i and j; k medoids $c_1, \ldots, c_k \in E$

Result: Clusters $G_1, \ldots, G_k \subset E$

repeat
    forall *item* $i \in E$ **do**
        Assign i to the closest medoid, creating clusters $G_1, \ldots, G_k \subset E$
    forall *medoid* c_j **do**
        forall *item* $i \in E$ **do**
            Compute the improvement (or the lost) of a solution where c_j is moved on item i
    if *A strictly positive improvement is found* **then**
        Move the medoid on the item inducing the largest improvement
until *no strict improvement is found*

The PAM algorithm has complexity in $\Omega(k \cdot n^2)$. The computation of the cost of the new configuration on Line 6 of the algorithm requires an effort proportional to n. Indeed, it is necessary to check, for each element not associated with c_j, if the new medoid i is closer to the current one. For the elements previously associated with the medoid c_j, the best medoid is either the second best of the previous configuration, which can be pre-calculated and stored in Line 3, or the new medoid i tried.

The number of repetitions of the loop ending in Line 9 is difficult to assess. However, we observe a relatively low number in practice, depending on k more or less linearly (there is a high probability that each center will be repositioned once) and a sub-linear growth with n. If we want a number of clusters k proportional to n (for instance, if we want to decompose the set E into clusters comprising a fixed number of elements, on average), the complexity of Algorithm 2.7 is higher than n^4. Thus, the algorithm is unusable as soon as the number of elements exceeds a few thousand.

2.7.2 *k-Means*

In case the items are vectors of real numbers and the measurement of the dissimilarity corresponds to the square of the distance (l_2 norm), the point μ that minimizes the homogeneity of the star criterion associated with a group G is the arithmetic average of elements of G (the center of gravity). The *k-means* heuristic Algorithm 2.8 is probably the best known algorithm for clustering.

Algorithm 2.8: (k-means) Local search improvement method for clustering items of $\mathbb{R}^d$ into k groups. The dissimilarity measure is the l_2 norm

Input: Set E of items in $\mathbb{R}^d$ with l_2 norm measuring the dissimilarity between items; k centers $c_1, \ldots, c_k \in \mathbb{R}^d$
Result: Clusters $G_1, \ldots, G_k \subset E$
1 **repeat**
2 **forall** *item* $i \in E$ **do**
3 Assign each item $i \in E$ to its nearest center, creating clusters $G_1, \ldots, G_k \subset E$
4 **forall** $j \in 1, \ldots, k$ **do**
5 $c_j =$ gravity center of G_j
6 **until** *no center has moved*

Similar to the PAM algorithm, this is a local search improvement method. It starts with centers already placed. Frequently, the centers are randomly positioned. It alternates an assignment step of the item to their nearest center (Line 3) and an optimal repositioning step of the centers (Line 5). The algorithm stops when all items are optimally assigned and all centers optimally positioned considering their assigned items. This algorithm is relatively fast, in $\Omega(k \cdot n)$. Unluckily, it is extremely sensitive to the initial center positions as well as isolated items. If the item dispersion is high, another dissimilarity measure should be used. For instance, the centers can be optimally repositioned considering the ordinary distance (l_1 norm) at Line 5. This variant is the Weber's problem. By replacing the center c_j on the medoid item of group G_j at Line 5, a faster variant of Algorithm 2.7 is obtained.

2.8 Graph Coloring

Coloring the edges or the vertices of a graph allows us to mentally represent problems where incompatible items must be separated. Two compatible items can receive the same "color" while they must be colored differently if they are incompatible. Therefore, a color represents a class of compatible elements. In the edge coloring, two edges having a common incident vertex must receive different colors. In the vertex coloring, two adjacent vertices must receive different colors.

The edge coloring can be transformed into a vertex coloring in the line graph. Building the line graph $L(G)$ from the graph G is illustrated in Fig. 2.12.

The vertex coloring problem is to find the chromatic index of the graph, that is to say, to minimize the number of colors of a feasible coloring. This problem is NP-hard in the general case. However, the edge coloring of a bipartite graph can be solved in polynomial time.

2.8.1 Edge Coloring of a Bipartite Graph

It is clear that the vertices of a bipartite graph can be colored with two colors. It is a bit more complicated to color the edges of a bipartite graph. But we can find an

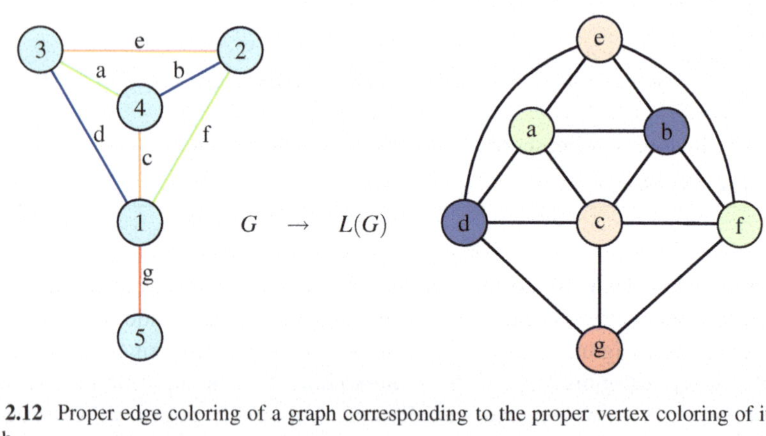

Fig. 2.12 Proper edge coloring of a graph corresponding to the proper vertex coloring of its line graph

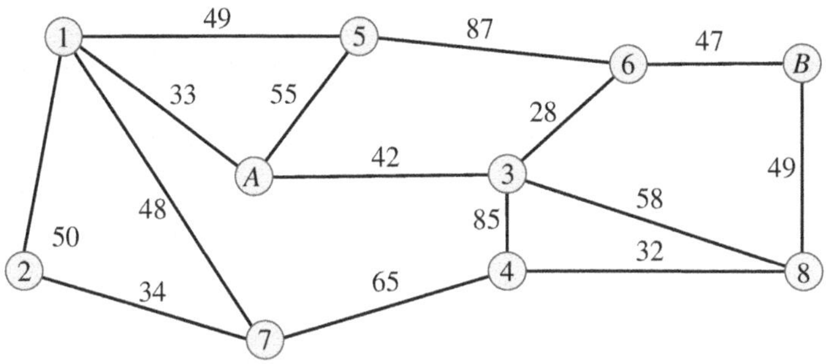

Fig. 2.13 Maximum load problem in a road network

optimal coloring in polynomial time. For this, we begin by completing the bipartite graph $G = (V = X \cup Y, E)$ by adding vertices to the smallest subset X or Y so that they contain the same number of vertices. While maintaining it bipartite, edges are added to the graph so that all its vertices have the same degree. This degree equals the largest of a vertex of G.

Let us call G' the bipartite graph so obtained. A perfect matching can be found in G' by solving a maximum flow problem (see Sect. 2.5). The edges of this matching can use color number 1. Then, the edges of this matching are removed from G' to obtain the graph G''. The last has the same properties as G': it is bipartite, both subsets containing the same number of vertices and all their degree being the same. So, a perfect matching can be found in G'', the edges of this matching receiving the color number 2. The process is iterated until no edge remains in the graph. The coloring so obtained is optimal for G (and for G') because the number of colors used is equals to the vertex of G with the highest degree. See also Problem 3.3.

Problems

2.1 Connecting Points

A set V of points on the Euclidean plane must be connected. How to proceed to minimize the total length of the connections? Application: consider the 3 points $(0, 0)$, $(30, 57)$, and $(66, 0)$.

2.2 Accessibility by Lorries

In the road network of Fig. 2.13, the maximum load (in tons) is given for each edge. What is the weight of the heaviest lorry that can travel from A to B?

2.3 Network Reliability

Figure 2.14 gives a communication network where connections are subject to breakdowns. The reliability of the connections is given for each edge. How should

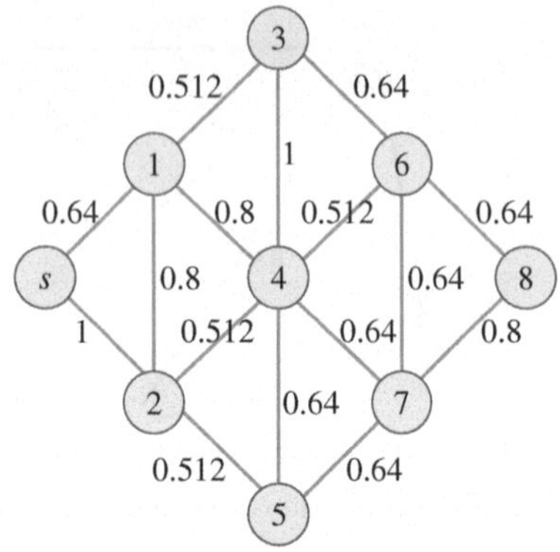

Fig. 2.14 Reliability in a network

we transmit a message from the vertex s to all the others with the highest possible reliability?

2.4 Ford and Fulkerson Algorithm Degeneracy
Show that the Ford and Fulkerson algorithm for finding a maximum flow is not polynomial.

2.5 TSP Permutation Model
Model the TSP under the form of finding an optimal permutation.

2.6 PAM and k-Means Implementation
Implement Algorithms 2.7 and 2.8 by initializing the k medoids or the k centers with the first k items. Investigate both methods on randomly generated problems in the unit square with $n = 100, 200, 400, 1000, 2000$ items and $k = 5, 10, 20, 50, n/20$ centers. Estimate the empirical complexity of the algorithms. Compare the quality of the solutions obtained by Algorithm 2.8 when the k centers are initially placed on the medoids found by Algorithm 2.7 rather than randomly choosing them (with the k first items).

2.7 Optimality Criterion
Prove that the schedule given at the bottom of Fig. 2.4 is optimal.

2.8 Flowshop Makespan Evaluation
Knowing the processing time t_{ij} of the object i on the machine j, how to evaluate the earliest ending time f_{ij} and the latest starting time d_{ij} for a permutation flowshop? The jobs are processed in an order given by the permutation p.

2.9 Transforming the Knapsack Problem into the Generalized Assignment
Knowing that the knapsack problem is NP-hard, show that the generalized assignment problem is also NP-hard.

References

1. Jarník, V.: O jistém problému minimálním. (Z dopisu panu O. Borůvkovi [On a certain problem of minimization (from a letter to O. Borůvka)], in Czech. Práce moravské přírodovědecké společnosti. **6**(4), 57–63 (1930). http://dml.cz/dmlcz/500726
2. Jarník, V.: O minimálních grafech, obsahujících n daných bodů [On minimal graphs containing n given points], in Czech. Časopis pro Pěstování Matematiky a Fysiky. **63**(8) 223–235 (1934). https://doi.org/10.21136/CPMF.1934.122548
3. Kaufman, L., Rousseeuw, P.J.: Clustering by means of Medoids. In: Dodge, Y. (ed.) Statistical Data Analysis Based on the L_1-Norm and Related Methods, pp. 405–416. North-Holland, Amsterdam (1987)
4. Martello, S., Toth, P.: Knapsack Problems — Algorithms and Computer Implementations. Wiley, Chichester (1990)
5. Voigt (ed.): Der Handlungsreisende wie er sein soll und was er zu thun hat, um Aufträge zu erhalten und eines glücklichen Erfolgs in seinen Geschäften gewiß zu sein ; Mit einem Titelkupf. Voigt, Ilmenau (1832)

Chapter 3
Problem Modeling

In all fields, it is essential to choose a good model for the problem to be addressed. Indeed, the best solution method will be useless if it is given inappropriate data or constraints. Let us illustrate this on the Steiner tree problem. Two simple modelings may naturally be imagined:

- **Steiner nodes to retain** A solution can be represented by the Steiner node to belong to the tree; knowing these nodes, the tree is constructed by the application of the Prim or the Kruskal algorithm.
- **Edges to retain** A solution can equally be represented by a set of edges; these edges must produce a connected graph containing all terminals.

It is not possible to determine a priori which model is the best. It really depends on the type of algorithm that will be developed to solve the problem. For example, the first model might be better suited to a constructive algorithm, while the second might be better suited to a local search.

The first part of this chapter gives various modeling examples for the graph coloring problem. It presents some techniques to transform the objective and the constraints of an optimization problem in order to obtain a model facilitating the design of solving algorithms.

Sometimes, there is not just one clear objective to optimize, but several. The next part of this chapter introduces some concepts of multi-objective optimization.

When faced with a new problem, it is not necessarily obvious how to find a good model. Sometimes, the "new" problem may even be a classic one that has not been recognized. The last part of this chapter gives some examples of practical applications that can be modeled as combinatorial problems presented in the previous chapter.

É. D. Taillard, *Design of Heuristic Algorithms for Hard Optimization*, Graduate Texts in Operations Research, https://doi.org/10.1007/978-3-031-13714-3_3

3.1 Objective Function and Fitness Function

To ensure the problem has been understood, it is necessary to formally express or model its core. A problem can be modeled in various ways. Next, the solving methods must be adapted to the chosen model. This section presents various models for the vertex coloring problem. These models are illustrated with a small graph. Figure 3.1 gives an optimal coloring of this graph.

In Sect. 1.1, we have seen that this problem could be modeled by a satisfiability problem. If a coloring with a minimum number of colors is wanted, a series of satisfiability problems can be solved. Unless working on small graphs and having an efficient satisfiability solver, this approach is hardly practicable. Another modeling, presented in the same section, consists in formulating an integer linear program. The objective (1.4) is to directly minimize the number of colors used. In general terms, a combinatorial optimization problem can be formulated as:

$$\text{Optimize}: \quad f(s) \tag{3.1}$$

$$\text{Subject to}: \quad s \in S \tag{3.2}$$

The correspondence between this general model and the linear program presented in Sect. 1.1.1 is as follows: the objective (3.1) is to minimize (1.4), which is equivalent to minimizing the highest color index. Constraint (3.2) summarizes Constraints (1.5), (1.6), (1.7), and (1.8).

The graph coloring problem can be expressed in a less formal way as:

$$\text{Minimize}: \quad c = f_1(s) \tag{3.3}$$

$$\text{Subject to}: \quad \text{Two adjacent vertices have different colors} \tag{3.4}$$

$$\text{and}: \quad \text{Number of colors used by } s \leqslant c \tag{3.5}$$

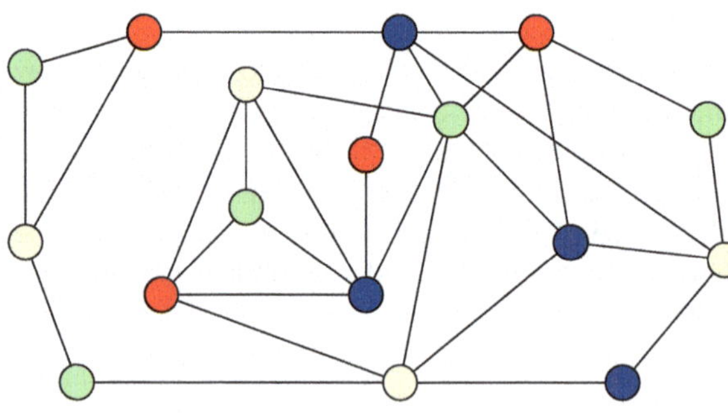

Fig. 3.1 Coloring the vertices of a graph with a minimum of color. This coloring is the best possible since the graph contains a clique of four vertices

Here, function f_1 returns the number of the highest color used. Figure 3.2 provides a feasible coloring of a graph that is not optimal. For the solution given in this figure, we have $f_1 = 5$. For those given in Fig. 3.1, we have $f_1 = 4$.

Another graph coloring model is less intuitive. It consists in directing the edges of the graph without creating a circuit. The objective is to minimize the length of the longest path in the directed graph:

$$\begin{aligned} \text{Minimize :} \quad & f_2(s) = \text{Longest path in } G \text{ oriented} \\ \text{Subject to :} \quad & \text{The edge orienting of } G \text{ has no circuit} \end{aligned} \tag{3.6}$$

Once such a directed graph is obtained, a feasible solution can be easily found. The vertices without predecessor receive the color number 1. They cannot be connected by an arc, so there is no constraint violation.

These vertices can be removed from the graph before assigning the color number 2 to those staying without predecessor and so on. Hence, the number of colors obtained is one more than the length of the longest path. The coloring of a graph with this model is illustrated in Fig. 3.3.

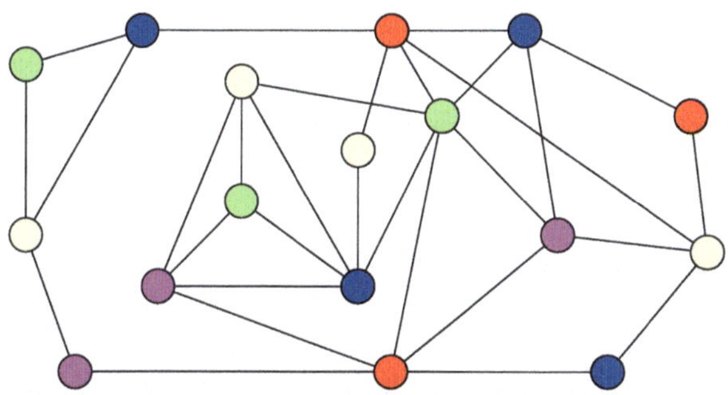

Fig. 3.2 Graph colored with too many colors. A number of changes are required to remove the unnecessary color

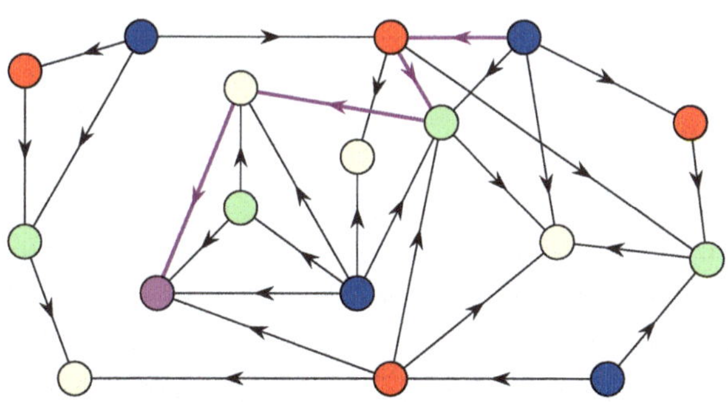

Fig. 3.3 Graph coloring obtained by directing the edges without circuit. One of the longest paths is indicated in bold. For this solution, $f_2 = 4$ corresponds to a 5-coloring

Minimizing a maximum—or maximizing a minimum—as (3.3) or (3.6) is generally not a good way for easily discovering or improving solutions satisfying all constraints. In the context of local search, such a modeling contains very large plateaus (see Sect. 5.1.3, Fig. 5.2).

What if a solution that has been discovered uses one color more than the optimum, as shown in Fig. 3.2? Is there just a vertex that uses the extra color, or are there many? The objective function is often replaced by a *fitness function*, easier to optimize, for instance, by including constraints that have been relaxed.

3.1.1 Lagrangian Relaxation

A problem that can be put in the form:

$$\begin{aligned} &\text{Minimize} \quad f(s) \\ &\text{Subject to:} \quad s \in S \quad \text{Easy constraint} \\ &\qquad \text{and: } g(s) \leqslant 0 \text{ Make the problem hard} \end{aligned}$$

could be modeled :

$$\begin{aligned} &\text{Minimize } f(s) + \lambda \cdot max(g(s), 0) \text{ with } \lambda \text{being a parameter} \\ &\text{Subject to:} \qquad s \in S \end{aligned}$$

If λ is large enough, both models have the same optimal solution.

3.1.1.1 Lagrangian Relaxation for the Vertex Coloring Problem

For the vertex coloring problem, Constraint (3.4) can be relaxed. A Lagrangian relaxation of the problem is:

$$\begin{aligned} &\text{Minimize :} \quad f_3(s) = c + \lambda \cdot \text{ Number of violations of (3.4)} \\ &\text{Subject to :} \quad (3.5) \end{aligned} \tag{3.7}$$

For a sufficiently large λ value (for instance, by setting λ = chromatic number), a solution optimizing (3.7) is also optimal for (3.3). In that manner, a triangle colored with a single color has a fitness of $1 + 3\lambda$. The fitness is $2 + \lambda$ with two colors, and the optimal coloring has a fitness of 3. For $0 < \lambda < 1/2$, the optimum solution of f_3 has one color; for $1/2 < \lambda < 1$, it has two colors; and for $\lambda > 1$, the optimum is a feasible solution with three colors. For instance, the solution of Fig. 3.1 has a fitness of $f_3 = 4$, that of Fig. 3.2 is $f_3 = 5$, and that of Fig. 3.4 is $f_3 = 4 + 2\lambda$.

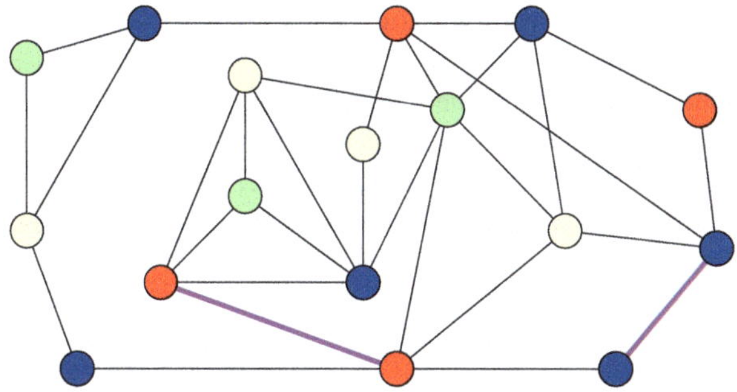

Fig. 3.4 Unfeasible coloring of a graph with a given number of colors. For this solution, we have $f_3 = 4 + 2\lambda$. For those of Fig. 3.1, we have $f_3 = 4$. For those of Fig. 3.2, we have $f_3 = 5$

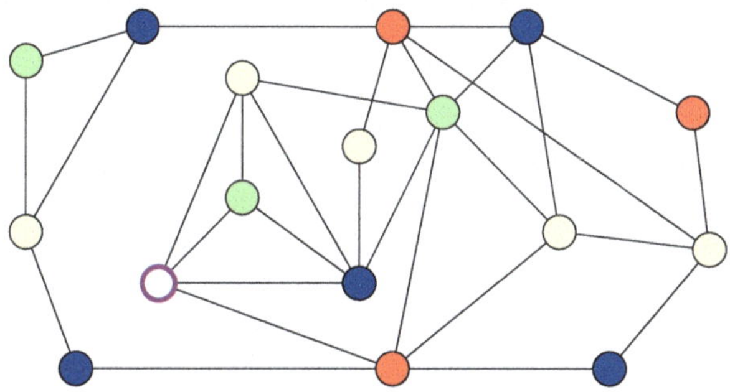

Fig. 3.5 Partial coloring of a graph with a given number of colors. For this solution, we have $f_4 = 1$

Instead of relaxing the constraint that no two adjacent vertices receive the same color, we can relax the constraint that each vertex be colored:

$$\begin{aligned} \text{Minimize} &: \quad f_4(s) = \text{Number of uncolored vertices} \\ \text{Subject to:} &: \quad (3.4) \end{aligned} \tag{3.8}$$

A partial coloring of a graph with $f_4 = 1$ is given in Fig. 3.5.

Generally, the value of the multiplier λ associated with a relaxed constraint placed in the fitness function is modulated according to the success of the search: if a feasible solution is discovered, the value of λ is diminished. Conversely, if all generated solutions are unfeasible, the value of λ is increased. Let us illustrate this for the TSP.

3.1.1.2 Lagrangian Relaxation for the TSP

A *1-tree* of minimum weight in a network with vertices $1, 2, \dots, n$ is a minimum spanning tree on the vertices $2, \dots, n$ plus two edges adjacent to vertex 1 with the lowest weights. Node 1 carries out a particular role, hence the term *1-tree*. We can reformulate the TSP by specifying that one seeks a 1-tree of minimum weight with the constraint imposing a degree of two for every vertex.

$$\begin{aligned} \min z &= \textstyle\sum_{(i,j)\in H} d_{ij}x_{ij} \\ \text{Subject to: } &\textstyle\sum_{j=1}^{n} x_{ij} = 2 \quad (i = 1, \dots, n) \\ &\text{and } H \text{ is a 1-tree} \end{aligned} \tag{3.9}$$

By relaxing Constraints (3.9) on the degree of each vertex and including them with Lagrange multipliers in the objective, we get:

$$\begin{aligned} &\min z(\lambda) = \textstyle\sum_{(i,j)\in H} d_{ij}x_{ij} + \sum_{i=1}^{n} \lambda_i(\sum_{j=1}^{n} x_{ij} - 2) \\ &\text{Subject to: } H \text{ is a 1-tree} \end{aligned} \tag{3.10}$$

For fixed $\lambda_i (i = 1 \dots n)$, the problem reduces to the construction of a 1-tree of minimum weight in a network where the weight of the edge (i, j) is shifted to $d_{ij} + \lambda_i + \lambda_j$. With these modified weights, the length of a tour is the same as with unmodified weights, but increased by $2 \cdot \sum_i \lambda_i$. Indeed, it is necessary to enter once in each vertex i and come out once, having to "pay" a λ_i penalty twice. Therefore, $z(\lambda)$ provides a lower bound to the length of the optimal tour. The value of this bound can be improved by finding λ_i maximizing $z(\lambda)$.

Figure 3.6 illustrates various 1-tree that can be obtained by modifying the λ values for a small TSP instance.

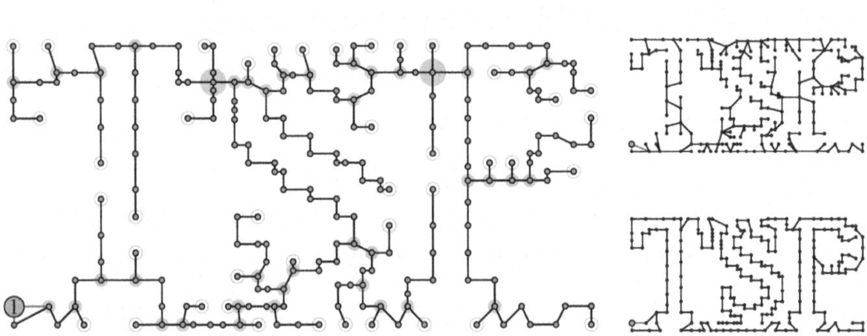

Fig. 3.6 Left: 1-tree on the TSP instance *tsp225* obtained with all λ values set to 0. The size of filled disks is proportional to the value to add to the λ associated with penalized vertices, and circles are proportional to the value to remove. Top right: the 1-tree obtained with the first modification of the λ. Bottom right: the 1-tree obtained after iterating the process. Only 12 vertices have a degree different from 2, and the length of this 1-tree is about 7.5% above the length of the initial 1-tree and 1.1% below the optimum tour length

3.1.2 Hierarchical Objectives

Rather than introducing several constraints with penalties in the fitness function, another possibility is to consider hierarchical objectives. For the graph coloring problem, a primary objective counts the number of colors used. The value of this objective is chosen and fixed. A secondary objective measures the number of constraints violation. Once a feasible solution is found, one can try reducing the number of colors before starting again. If no feasible solution is achieved, the number of colors of the primary objective is increased. Proceeding like this allows not completely losing the work done so far.

The modeling choice has a significant influence on the problem-solving capacity of a given method. Therefore, it is worth paying close attention to the problem analysis phase. For instance, if a timetable problem is modeled by a graph to color, the user might not necessarily be interested in the timetable using the smallest possible time slots. This last objective is minimized by solving a standard graph coloring problem. The user might just specify the maximum number of time slots available. Then any timetable using no more than these time slots could be fine. In this case, a fitness function of type f_3 (with λ close to 0 and c fixed to the maximum number of time slots desired) or f_4 would certainly be more convenient than f_1 or f_2.

Relaxing constraints by including a penalty for their violations in the fitness function can only be considered for a relatively small number of constraints. However, proceeding like this is very common when we are dealing with "soft" constraints. The last corresponds rather to preferences than to strict constraints. In this case, using a model with hierarchical objectives may be a good option.

A more flexible and frequently used approach is to consider several objectives simultaneously and to leave the user the choice of a compromise—prefer a solution good for an objective rather than another. Such an approach implies the methods to be able to propose various solutions rather than a single one optimizing a unique objective.

3.2 Multi-Objective Optimization

Route planning is a typical multi-objective optimization example. To go from a point A to a point B, we have to use a given transportation network. This constitutes the constraints of the problem. We want to minimize the travel time of the journey, minimize the energy consumption, and maximize the pleasure of the journey.

These objectives are generally antagonists. The same route can be done by reducing the energy consumption at the cost of an increased travel time. The user will choose an effective route on a more subjective basis: for instance, by "slightly" increasing duration, the energy consumption "sensibly" decreases, and the landscape is "picturesque."

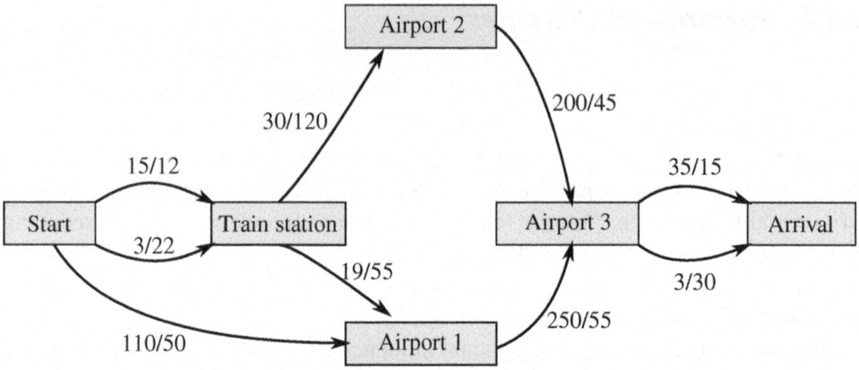

Fig. 3.7 Example of a multi-objective problem: To travel from a departure place to a destination, several routes can be selected. Each trip has a given cost and duration, indicated next to the arcs. From the departure place, we can either go to the train station by bus, or by taxi, or go directly to the nearest airport by taxi. From the train station, we can reach either the nearest airport or another airport. The last is a little further away but better serviced and flights are more competitive. Then we fly up to the airport the closest to the destination, where we can go to the final destination either by bus or by taxi

Figure 3.7 illustrates the case of someone who has to travel by air and who has the choice between several means of transportation to get to an airport and reach the final destination.

An optimization problem with K goals can formulate generally by:

$$\begin{aligned} \text{"Minimize"}: \quad & \vec{f}(s) = (f_1(s), \ldots, f_K(s)) \\ \text{Subject to}: \quad & s \in S \end{aligned} \tag{3.11}$$

This formulation assumes that one seeks to minimize each objective. This does not constitute a loss of generality. Indeed, it is equivalent to maximizing or minimizing the opposite. It is said that a solution s_1 *dominates* a solution s_2 if s_1 is better than s_2 on at least one objective and at least as good as s_2 on the other objectives.

The purpose of multi-objective optimization is therefore to exhibit all non-dominated solutions. These solutions are qualified as *efficient* or *Pareto-optimal*. Representing a solution where each axis represents the value of each objective, the solutions that are on the convex envelope of the *Pareto frontier* are called the *supported solutions*.

Figure 3.8 provides all the solutions to the problem instances given in Fig. 3.7 in a cost/time diagram. We see in this figure that there are efficient solutions not located on the convex hull.

The ultimate choice of a solution to the problem is left to the decision-maker. This choice is subjective, based, for instance, on political, ethical, or other considerations that cannot be reasonably quantified and therefore introduced neither in the objectives nor in the constraints.

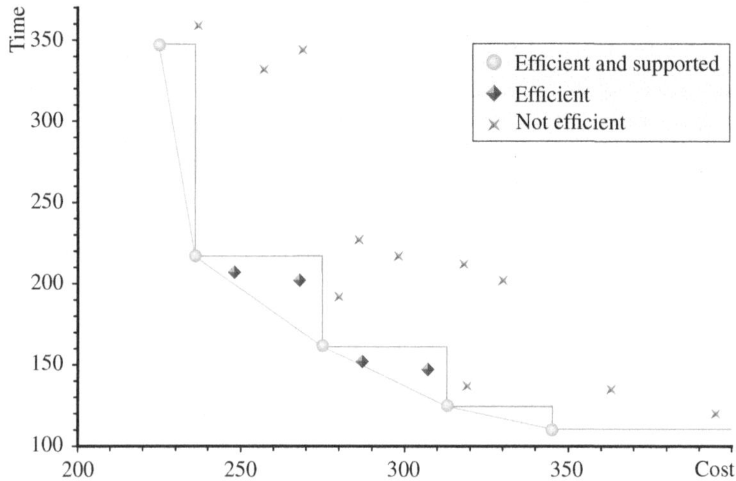

Fig. 3.8 Representation of the solutions of the problem of Fig. 3.7 in a cost/time diagram. The nine efficient solutions are highlighted. Only supported solutions may be discovered by an exact scalar method

3.2.1 Scalarizing

A technique for generating various solutions to a multi-objective problem is to modify the last in a single-objective problem. The parameters of this problem are as many as the objectives of the multi-objective problem. Thus, the multi-Objective Problem (3.11) turns into a single-objective problem with parameters $w_1, \ldots, w_K$.

$$\begin{aligned} \text{Minimize} &: \sum_{i=1}^{K} w_i \cdot f_i(s) \\ \text{Subject to} &: s \in S \end{aligned} \tag{3.12}$$

This technique is known as *linear scalarization*. Let us suppose we have an exact method for solving Problem (3.12). The supported solutions can be discovered by varying the value of the weights w_i. For instance, by setting $w_i = 1$ for a given i and zero weights for the other objectives, the best possible solution for the ith criterion can be found. Once these K solutions are known, other supported solutions can be found, if any exists. Indeed, vectors (w_i) orthogonal to the hyperplane supported by the K objectives of these K solutions can be considered. By reiterating the process with this new solution, the proper set of supported solutions can be generated.

3.2.2 *Sub-goals to Reach*

The main issue with linear scalarization is that the unsupported efficient solutions are not achievable. The extreme case is that the only supported solutions are the K individually optimizing a single objective. Unsupported solutions can be extracted by fixing the minimum quality of a solution on one or more objectives. The idea is to include one or more constraints while removing the same number of objectives. If the solutions are constrained to get a value at most v_1 for the first objective, we have the problem with $d - 1$ objectives:

$$\begin{aligned} \text{Minimize :} \quad & \vec{f}(s) = (f_2(s), \ldots, f_K(s)) \\ \text{Subject to :} \quad & f_1(s) \leqslant v_1 \\ \text{with :} \quad & s \in S \end{aligned} \tag{3.13}$$

In the example of Fig. 3.8, imposing a maximum budget $v_1 = 250$, the solution of cost 248 for a time of 207 is found. The last is not supported.

3.3 Practical Applications Modeled as Classical Problems

Let us conclude this chapter by giving some examples of practical applications that can be modeled as academic problems.

3.3.1 *Traveling Salesman Problem Applications*

As we have taken the traveling salesman problem as the main thread of this book, we start by showing how to model in this form problems that have a priori nothing to do with performing tours.

3.3.1.1 Minimizing Unproductive Moves in 3D Printing

To produce parts in small quantities, especially to make prototypes directly with CAD software, additive manufacturing or 3D printing techniques are now used. One of these techniques is to extrude a thermoplastic filament which is deposited layer by layer and hardens immediately.

It is particularly useful to minimize the unproductive moves of the extrusion head when printing three-dimensional parts. To produce such a piece, the 3D model is sliced into layers of thickness depending on the diameter of the extrusion head. Each layer is then decomposed into a number of segments. If the extrusion head must start

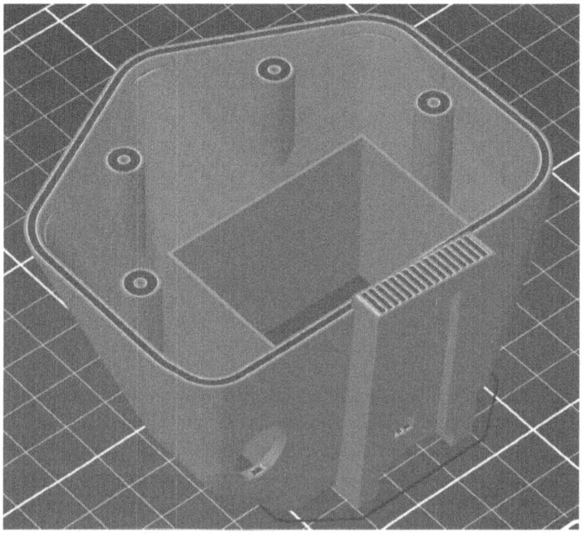

Fig. 3.9 Part of a bicycle lamp, viewed in the *PrusaSlicer* software

printing a segment at a different position than those of the previous segment, it must perform an on-air move. The total length traveled by the head is consistent: for the part shown in Fig. 3.9—about 10 cm wide, it represents nearly a kilometer. The unproductive moves can represent a relatively consequent fraction of all the moves. As a result, minimizing the latter allows a significant productivity gain.

The transformation principle of this problem into the traveling salesman is the following: a naive attempt would create a city per segment endpoint. However, a tour comprising only these cities does not necessarily provide a solution to the extrusion problem, as a segment might not be printed when visiting one of its endpoints.

Therefore, it is necessary to force the tour to visit the other endpoint of a segment immediately after visiting the first endpoint. For this purpose, a city is included in the middle of each segment. The standard TSP does not have constraints on the city visiting order. To ensure a good TSP tour corresponds to a feasible solution to the extrusion problem, the distance matrix must be adapted. To force printing a segment, the distance between the cities' endpoints and the middle city is zero. The head will truly take a while to print the segment, but this time is incompressible since the segment must be printed anyway.

To prevent certain moves of the head, a large distance M is associated with prohibited moves. In this manner, a proper tour will not include such moves of the head. The value of M should not be too large, to avoid numerical problems. A value about 100 times the size of the object to be printed may be suitable. This technique prevents connecting the middle city of a segment to all other cities but excepted both endpoints of the corresponding segment. Another constraint is to print all segments of a layer before printing the subsequent layer.

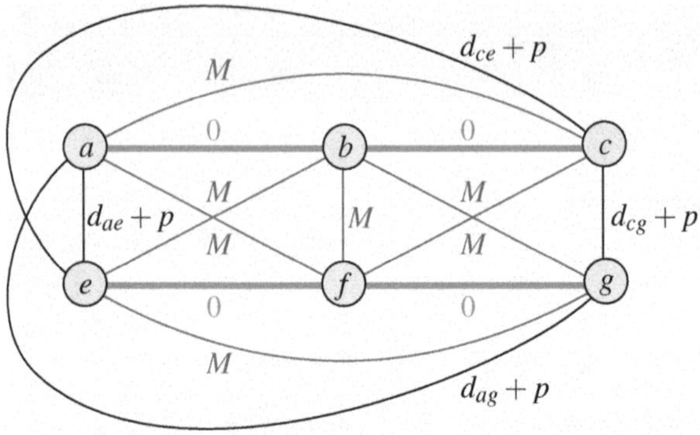

Fig. 3.10 Transformation of the problem of minimizing non-productive moves of a 3D printing into a TSP. Principle for assigning distances between six cities corresponding to segments $[a, c]$ and $[e, g]$ to print. The "cities" b and f are placed in the middle of the segments to force the extrusion head to go from one endpoint of a segment to the other one. The M value is relatively large (typically, 100 times the size of the object). The value of p depends on the respective layers of both segments: either 0 if they are in the same layer or $M/10$ if they are in adjacent layers or M if they are in different and non-adjacent layers

Indeed, it is no longer possible to extrude material below an already printed layer. Moreover, this would significantly complicate the management of the extrusion head. The last can collide with the material of an upper layer when printing a lower one. This can be prevented in the traveling salesman model by a technique similar to that presented above. Figure 3.10 illustrates how to build the distance matrix for printing two segments.

The distance between two endpoints of segments is penalized by a value p depending on the segment layers. If both segments belong to the same layer, the penalty is zero. Else, if the segments are in adjacent layers, we can set $p = M/10$. Thus, a proper tour goes only once from a layer to the next one. The length of a good tour corresponds to that of the unproductive moves plus $M/10$ times the number of layers to print. Otherwise, if the segments are neither in the same layer nor in adjacent layers, the penalty is set to $p = M$. Finally, two cities corresponding to the initial and ultimate positions of the extrusion head are added to complete the model.

It should be noted that traveling salesman models for minimizing unproductive moves can lead to large size instances. The part illustrated in Fig. 3.9 has a few hundred thousand segments. Figure 3.11 illustrates the productivity gain that can be obtained by optimizing unproductive moves. On the one hand, we have the moves proposed by the *PrusaSlicer* software for just one layer. On the other hand are the optimized moves obtained with a traveling salesman model. The length of the unproductive moves can be divided by about 9 for this layer.

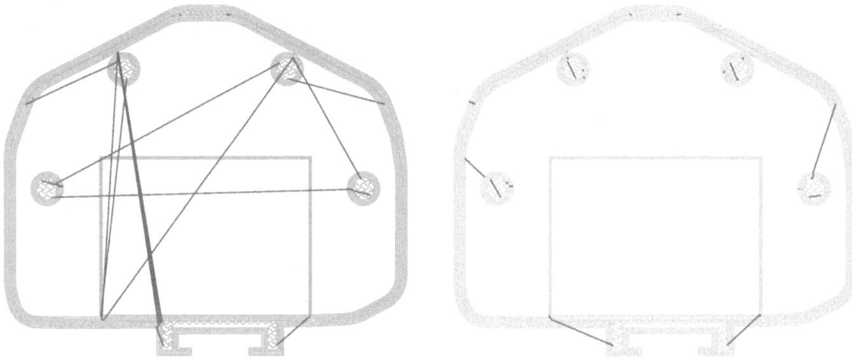

Fig. 3.11 On the left, moves of the extrusion head, as produced by the *PrusaSlicer* software. Darker segments correspond to unproductive moves. The total length of these moves is about 740.8 mm. On the right, optimized moves using a traveling salesman model. The length of non-productive moves is about 84.5 mm

3.3.1.2 Scheduling Coloring Workshop

One of the simplest scheduling problems is that of painting objects. The unique resource is the machine coloring the objects. Each task has only one operation: color the object i $(i = 1, \ldots, n)$; the duration of the operation is t_i. After coloring the object i, the machine must be cleaned to correctly color the next, j; this set-up time is s_{ij}. Note that, generally, $s_{ij} \neq s_{ji}$: indeed, dark pigments in a pastel color are more visible than the opposite. After coloring all the objects, the machine must be cleaned to be ready for the next day; the duration of this operation is r. The goal is to find the best coloring order of the objects. Hence, we look for a permutation p of the n objects minimizing $\sum_{i=1}^{n-1} (t_{p_i} + s_{p_i p_{i+1}}) + t_{p_n} + r$. This scheduling problem with set-up time can be reduced to a traveling salesman instance with $n + 1$ cities. For proving this, let $w_{ij} = t_i + s_{ij}(i, j = 1, \ldots, n)$, $w_{i0} = r$, and $w_{0i} = 0, (i = 1, \ldots, n)$. We can verify that the shortest tour on the cities $0, \ldots, n$ provides the optimal order to color the objects. The "object" 0 represents the beginning of a workday.

3.3.2 Linear Assignment Modeled by Minimum Cost Flow

The linear assignment mathematical model given in Sect. 2.5.1 is both concise and rigorous. However, it does not indicate how to solve a problem instance. Using general integer linear programming solvers might be inappropriate—unless the solver automatically detects assignment constraints (2.34) and (2.35) and incorporates an ad hoc algorithm to process them. This is frequently the case.

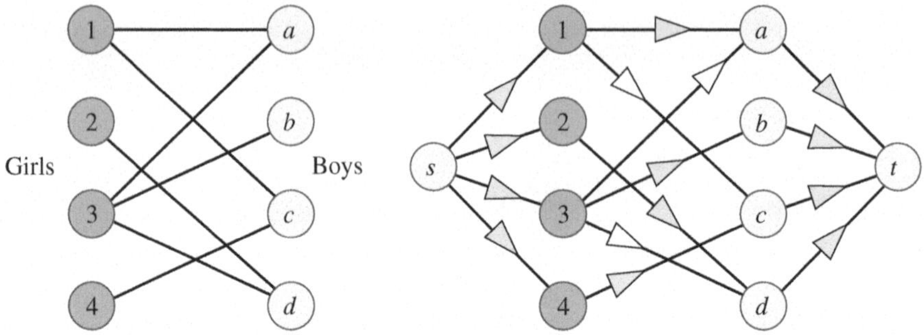

Fig. 3.12 Matching problem: how many mixed couples can be formed between girls and boys? A compatible matching is represented by an edge in a bipartite graph. The problem can be modeled by searching for a maximum possible flow from a vertex s to a vertex t in a network

The maximum matching problem can be treated as a special linear assignment problem: if i can be matched with u, then we can set a cost of 0 to the edge; otherwise, a positive cost is set. If an assignment of minimum cost is found, the last uses as few edges with positive cost and as many edges with 0 cost as possible. The maximum matching problem can be modeled with a maximum flow in a bipartite network. Figure 3.12 illustrates how a matching problem can be solved with a flow in a network.

Similarly, the linear assignment problem can be modeled by a minimum cost flow. A bipartite network $R = (I \cup U \cup \{s\} \cup \{t\}, E, w, c)$ is built similarly to the matching problem presented in Fig. 3.12. Every pair of nodes $(i \in I, u \in U)$ is connected by an arc (i, u) with capacity $w(i, u) = 1$ and cost $c(i, u)$. An arc (s, i) of capacity $w(s, i) = 1$ with cost 0 connects s to each node $i \in I$. An arc (u, t) of capacity $w(u, t) = 1$ with cost 0 connects each node $u \in U$ to t.

Finding an optimum cost flow in R allows finding the optimal assignment in polynomial time, for instance, with Algorithm 2.6.

More efficient algorithms have been designed for the linear assignment problem. This leads us to make a comment on the integer linear program presented in Sect. 2.5.1. A peculiarity of the constraint matrix is that it contains only 0s and 1s. It can be proved that the adjacency matrix of a bipartite graph is *totally unimodular*. This means that the determinant of any square sub-matrix is either -1 or 1 or 0. Hence, the integrality constraints (2.36) can be omitted. Therefore, the standard linear program provides an integer optimal solution.

3.3.3 Map Labeling Modeled by Stable Set

An application of the maximum weight stable set appears for map labeling. When one wishes to associate information with objects drawn on a plan, the problem is to choose the label position so that they do not overlap. Figure 3.13 illustrates a tiny problem of labeling three cities.

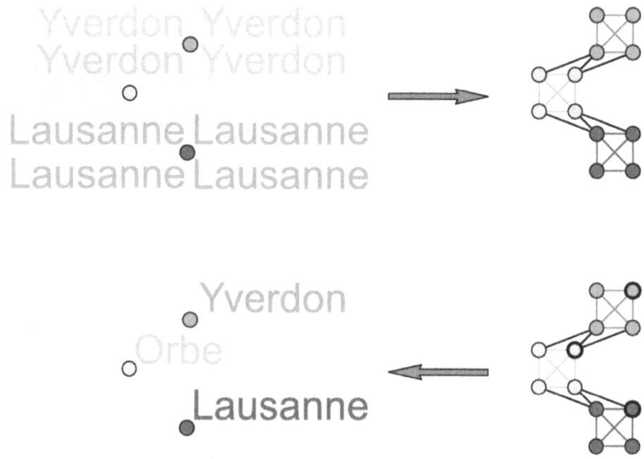

Fig. 3.13 Map labeling problem: The name of three objects must be placed on a map. The texts should not overlap to maintain readability. For this instance, four possibilities are considered for the label position of each object. In terms of a graph, this problem can be modeled by a maximum stable set. The vertices of the stable correspond to the chosen label positions. Here, the name of each city can be placed at the top, right corner

This problem can be transformed into the maximum stable set as follows: A vertex is created for each potential label position. The set of vertices corresponding to the same label is connected in a clique. Indeed, there should be only one label per object. The vertices corresponding to overlapping labels are also connected with an edge. Hence, a stable set corresponds to label positions without overlap. To display as many labels as possible on the map, a maximum stable is searched in the graph.

In practice, not all positions are equivalent. Indeed, according to the language and the culture, there are preferred positions. For instance, in Western countries, one prefers to place the names at the top, right corner rather than at the bottom, left one. Preferences can be modeled as weights associated with positions. The map labeling problem then consists in finding a *maximum weight stable set.*

Other problems can be modeled in exactly the same way. First is the berth allocation for docking ships. Translated in terms of labeling, a rectangular label is associated with each ship. The label width is the expected duration of the stopover, and the label height is the length of the ship. The possible positions for this label are determined by the ship's arrival time and by the dock locations that can accommodate the boat.

Another application of this problem is the flight levels allocation for commercial aircraft. Knowing the expected departure times of each aircraft and the routes they take, the potential areas of collision between aircraft are first determined. The size of these zones depends on the uncertainty of the actual departure times and the effective routes followed. A label will therefore correspond to a zone. The possible positions of the labels are the various flight levels that the aircraft could use.

Problems

3.1 Assigning Projects to Students

Four students (A, B, C, D) must choose among a set of four semester projects $(1, 2, 3, 4)$. Each student makes a grade prediction for each project. What choices students should make to maximize grade point average?

		Project			
		1	2	3	4
	A	6.0	5.0	5.8	5.5
Student	B	6.0	5.5	4.5	4.8
	C	4.5	6.0	5.4	4.0
	D	5.5	4.5	5.0	3.8

3.2 Placing Production Units

A company has three existing production units 1, 2, and 3 and wants to open three new units, 4, 5, and 6. Three candidate locations a, b, and c are retained for the new units. Figure 3.14 illustrates the locations of the existing and new production units.

The parts produced must be transferred from an existing unit to a new unit using only the connections indicated in the figure. For instance, the distance between the unit 1 and the location b is $3 + 4 = 7$. The numbers of daily transfers between the existing and new units are given in Table 3.1.

Where to place these new production units to minimize the total transfer distance?

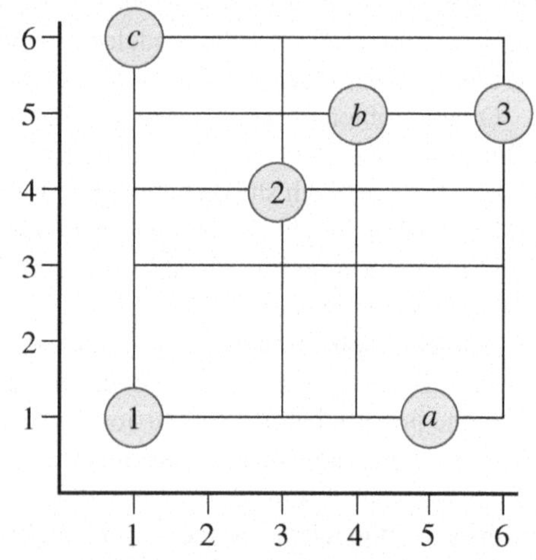

Fig. 3.14 Location of production units

Table 3.1 Number of daily transfers between existing and new units

	4	5	6
1	7	3	1
2	3	8	6
3	2	1	9

If parts are also transferred between new units, what kind of problem should we solve?

3.3 Oral Examination

Six students $(A, \ldots, F)$ undergo oral examinations for different modules $(1, \ldots, 5)$. The duration of each examination is 1 hour. How to build a timetable as short as possible, knowing that both students and teachers can have only one examination at a time?

		Student					
		A	*B*	*C*	*D*	*E*	*F*
	1	×		×	×		
	2	×	×		×	×	×
Module	3	×		×			×
	4			×		×	
	5		×	×			

3.4 Written Examination

The following table summarizes the student enrolments for written examinations of various modules. Each student can undergo one examination a day at most. All students to pass the same module are examined the same day. How many days, at minimum, are needed to organize the examination session?

		Student												
		A	*B*	*C*	*D*	*E*	*F*	*G*	*H*	*I*	*J*	*K*	*L*	*M*
	1	×	×							×		×		
	2	×			×									
	3	×					×				×	×		
	4		×						×			×		×
Module	5			×		×					×			×
	6			×				×					×	
	7			×					×					
	8				×		×	×		×	×			

3.5 QAP with More Positions than Items
How to adapt or change the QAP model when there are fewer elements to be placed (n) than positions (m)? Same question if there is a fixed cost c_{ir} for assigning the item i to the position r.

3.6 Mobile Phone Keyboard Layout
We wish to configure the keys of an old mobile phone. We want to place only the 26 lowercase letters as well as the space on the keys $0, 2, 3, \ldots, 9$. Up to four symbols can be placed per key.

For the language considered, g_{ij} represents the frequency of appearance of the symbol j after symbol i. To enter a symbol in position p on a key, it is necessary to press p times the key. This requires p time units. To switch from one key to another, the travel time is one unit. To enter two symbols located on the same key, we have to wait 6 time units. How to build a numerical instance of a QAP for this problem?

3.7 Graph Bipartition to QAP
The bipartition problem is to separate the vertices of a graph into two subsets X and Y of identical size (assuming an even number of nodes) so that the number of edges having an end in X and the other in Y are as low as possible. How to build a quadratic assignment instance for the graph bipartition?

3.8 TSP to QAP
How to build a quadratic assignment instance corresponding to a TSP instance?

3.9 Special Bipartition
We consider a set of cards numbered from 1 to 50. We want to split up the cards into two subsets. The sum of the numbers of the first should be 1170, and the product of the others should be 36,000. How to code a solution attempt to this problem? How to assess the quality of a solution attempt?

3.10 Magic Square
We want to create a magic square of order n. This square has $n \times n$ cells to be filled with the numbers of 1 to n^2. The sum of the numbers in each line, column, and diagonal must be $(n^3 + n)/2$. A magic square of order 4 is given below.

$$
\begin{array}{ccccccc}
 & & & & & & 34 \\
 & & & & & \nearrow & \\
4 & 14 & 15 & 1 & \rightarrow & 34 \\
9 & 7 & 6 & 12 & \rightarrow & 34 \\
5 & 11 & 10 & 8 & \rightarrow & 34 \\
16 & 2 & 3 & 13 & \rightarrow & 34 \\
\downarrow & \downarrow & \downarrow & \downarrow & \searrow & \\
34 & 34 & 34 & 34 & & 34
\end{array}
$$

How to code a solution attempt to this problem? How to assess the quality of a solution attempt?

3.11 Glass Plate Manufacturing

For producing glass plates, the molten glass passes on a chain of m machines in the order $1, 2, \ldots, m$. Depending on the desired features for each plate, the processing time on each machine differs. It is assumed there are n different plates to produce (in an order which can be decided by the chain manager). The processing time of the plate i on the machine j is t_{ij}. Additionally, when a machine has completed the processing of a plate, the latter must immediately switch to the next machine without waiting time; otherwise, the glass cools down. A machine only processes one plate at a time. The chain manager must determine in which order to produce the n plates to complete the production as quickly as possible. How to model this no-wait permutation flowshop problem as a TSP?

3.12 Optimal 1-Tree

Find values $\lambda_1 \ldots \lambda_5$ to assign to the five nodes of the network given in Fig. 2.2 such that:

- $\sum_{i=1}^{5} \lambda_i = 0$
- The weight of the 1-tree associated with these values is as high as possible

Part II
Basic Heuristic Techniques

This part introduces the building blocks of heuristics. First are the constructive methods. Then, once a solution is available, it can be improved with a local search. Finally, if either the problem is complex or the dataset is relatively large, decomposition methods can be used.

Chapter 4
Constructive Methods

Having ascertained that the problem to be solved is intractable and that the design of a heuristic is justified, the next step is to imagine how to construct a solution. This step is directly related to the problem modeling.

4.1 Systematic Enumeration

When we have to discover the best possible solution for a combinatorial optimization problem, the first idea that comes is to try to build all the solutions to the problem, evaluate their feasibility and quality, and return the best that satisfies all constraints. Clearly, this approach can solely be applied to problems of moderate size. Let us examine the example of a small knapsack instance in 0-1 variables with two constraints:

$$\begin{aligned} \max r &= 9x_1 + 5x_2 + 7x_3 + 3x_4 + x_5 \\ \text{Subject } & 4x_1 + 3x_2 + 5x_3 + 2x_4 + x_5 \leqslant 10 \\ \text{to}: & 4x_1 + 2x_2 + 3x_3 + 2x_4 + x_5 \leqslant 7 \\ & x_i \in \{0, 1\} (i = 1, \ldots, 5) \end{aligned} \tag{4.1}$$

To list all the solutions of this instance, an enumeration tree is constructed. The first node separates the solutions for which $x_1 = 0$ of those where $x_1 = 1$. The second level consists of the nodes separating $x_2 = 0$ and $x_2 = 1$, etc. Potentially, this problem has $2^5 = 32$ solutions, many of which are unfeasible, because of constraint violations. Formally, the first node generates two sub-problems that will be solved recursively. The first sub-problem is obtained by setting $x_1 = 0$ in (4.1):

É. D. Taillard, *Design of Heuristic Algorithms for Hard Optimization*, Graduate Texts in Operations Research, https://doi.org/10.1007/978-3-031-13714-3_4

$$\begin{aligned} \max r &= 0 + 5x_2 + 7x_3 + 3x_4 + x_5 \\ \text{Subject} \quad & 3x_2 + 5x_3 + 2x_4 + x_5 \leqslant 10 \\ \text{to :} \quad & 2x_2 + 3x_3 + 2x_4 + x_5 \leqslant 7 \\ & x_i \in \{0, 1\} (i = 2, \ldots, 5) \end{aligned}$$

The second sub-problem is obtained by setting $x_1 = 1$ in (4.1):

$$\begin{aligned} \max r &= 9 + 5x_2 + 7x_3 + 3x_4 + x_5 \\ \text{Subject} \quad & 3x_2 + 5x_3 + 2x_4 + x_5 \leqslant 6 \\ \text{to :} \quad & 2x_2 + 3x_3 + 2x_4 + x_5 \leqslant 3 \\ & x_i \in \{0, 1\} (i = 2, \ldots, 5) \end{aligned}$$

To avoid enumerating too many solutions, the tree can be pruned by noting that all branches arising from a node with a constraint violation will lead to unfeasible solutions. Indeed, for this problem instance, all constraint coefficients are non-negative. For instance, if the x_1, x_2, x_3 variables are already fixed to 1, both constraints are violated, and all the sub-problems that could be created from there will produce unfeasible solutions. Therefore, it is useless to develop this branch by trying to set values of the x_4 and x_5 variables.

Another way to prune the non-promising branches is to estimate by a short computation whether a sub-problem could lead to a better solution than the best found so far. This is the *branch and bound* method.

4.1.1 Branch and Bound

To quickly estimate whether a sub-problem may have a solution, and if the latter is promising, a technique is to *relax* one or more constraints. The optimal solution of the relaxed problem is not necessarily feasible for the initial one. However, few interesting properties can be deduced by solving the relaxed problem: If the latter has no solutions or its optimal solution is worse than the best feasible solution already found, then there is no need to develop the branch from this sub-problem. If the relaxed sub-problem contains an optimal solution feasible for the initial problem, then developing the branch is also unnecessary. In addition to the Lagrangian relaxation seen above (Sect. 3.1.1), several relaxation techniques are commonly used to simplify a sub-problem.

Variable integrality Imposing integer variables makes Problem (4.1) difficult. We can therefore remove this constraint and solve the problem:

$$\begin{aligned} \max S &= 9x_1 + 5x_2 + 7x_3 + 3x_4 + x_5 \\ \text{Subject } & 4x_1 + 3x_2 + 5x_3 + 2x_4 + x_5 \leqslant 10 \\ \text{to: } & 4x_1 + 2x_2 + 3x_3 + 2x_4 + x_5 \leqslant 7 \\ & 0 \leqslant x_i \leqslant 1 (i = 1, \ldots, 5) \end{aligned} \tag{4.2}$$

This linear problem can be solved efficiently in polynomial time. Its optimal solution is (0.5; 1; 1; 0; 0) with objective value of 16.5. Since it comprises a fractional value, this solution is not feasible for the initial problem. However, it informs us that there is no solution to Problem (4.1) whose value exceeds 16.5 (or even 16 since all the coefficients are integers). Therefore, if an oracle gives us the feasible solution (1; 0; 1; 0; 0) of value 16, we can deduce this solution to be optimal for the initial problem.

Constraint aggregation (surrogate constraint) A number of constraints are linearly combined to get another one. In our simple example, we get:

$$\begin{aligned} \max S &= 9x_1 + 5x_2 + 7x_3 + 3x_4 + x_5 \\ \text{Subject } & 8x_1 + 5x_2 + 8x_3 + 4x_4 + 2x_5 \leqslant 17 \\ \text{to: } & x_i \in \{0, 1\} (i = 1, \ldots, 5) \end{aligned} \tag{4.3}$$

This problem is a standard knapsack. It is easier to solve than the initial problem. The solution (1; 1; 0; 1; 0) is optimal for the relaxed Problem (4.3) but is not feasible for the initial problem because the second constraint is violated. As the relaxed problem is NP-hard, this approach may be problematic.

Combined relaxation Clearly, several types of relaxation can be combined, for instance, the aggregation of constraints and the integrality variables. For our small example, we get:

$$\begin{aligned} \max S &= 9x_1 + 5x_2 + 7x_3 + 3x_4 + x_5 \\ \text{Subject } & 8x_1 + 5x_2 + 8x_3 + 4x_4 + 2x_5 \leqslant 17 \\ \text{to: } & 0 \leqslant x_i \leqslant 1 (i = 1, \ldots, 5) \end{aligned} \tag{4.4}$$

This problem can be solved in $O(n \log n)$ as follows: the variables are sorted in the order of decreasing r_i/v_i values, where r_i represents the revenue of the object i and v_i its aggregated volume. In our example, the indices are already sorted. The objects are selected one after the other in this order until a new object would overcharge the knapsack. This leads to $x_1 = x_2 = 1$. The next object is split to completely fill the knapsack ($\Longrightarrow \ x_3 = 4/8$ for a total value of the knapsack $S = 9 + 5 + 7 \cdot 4/8 = 17,5$). Since all the coefficients are integers in our example, $S = \lfloor 17,5 \rfloor = 17$ is also a valid bound for the optimal value of the initial problem.

Algorithm 4.1 provides the general framework of the branch and bound method. Figure 4.1 shows a partial enumeration tree that can be obtained by solving the small problem instance (4.1). Three components should be specified by the user for implementing a complete algorithm.

Algorithm 4.1: Branch and bound framework for an objective to maximize. It is necessary to provide three methods: α for the management of the sub-problems to be solved (generally, a priority queue (based on a heuristic criterion) or a stack), a method β for evaluating the relaxation of the sub-problems, and a heuristic γ for choosing the next variable to separate for generating new sub-problems

```
Input: A problem with n variables x_1, ..., x_n, policy α for managing sub-problems,
       relaxation method β, branching method γ
Result: An optimal solution x* of value f*
f* ← −∞                                          // Value of best solution found
F ← ∅                                            // Set of fixed variables
L ← {x_1, ..., x_n}                              // Set of free variables
Q ← {(F, L)}                                     // Set of sub-problems to solve
while Q ≠ ∅ do
    Remove a problem P = (F, L) from Q according to policy α
    if P can potentially have feasible solutions with values already fixed in F then
        Compute a relaxation x of P with method β, modifying only variables x_k ∈ L
        if x is feasible for the initial problem and f* < f(x) then Store the improved
          solution
            x* ← x
            f* ← f(x)
        else if f(x) > f* then Expand the branch
            Choose x_k ∈ L according to policy γ
            forall possible value v of x_k do
                Q ← Q ∪ {(F ∪ {x_k = v}, L \ {x_k})}
        else No solution better than x* can be obtain
            Prune the branch
```

First, the management policy of the set Q of sub-problems awaiting treatment must be specified. If Q is managed as a queue, we have a breadth-first search. If Q is carried as a stack, we have a depth-first search. The last promotes a rapid discovery of a feasible solution to the initial problem.

A frequent choice is to manage Q as a priority queue. This implies computing an evaluation for each sub-problem. Ideally, this evaluation should be strongly correlated with the best feasible solution that can be obtained by developing the branch. A typical example is to use the value S of the relaxed problem. The choice of a management method for Q is frequently based on very empirical considerations.

The second component to be defined by the user is the relaxation technique. This is undoubtedly one of the most delicate points for designing an efficient branch and bound. This point strongly depends on both the problem to be solved and the numerical data.

The third choice left is how to separate the problem into sub-problems. A simple policy is to choose the smallest index variable or a non-integer variable in the

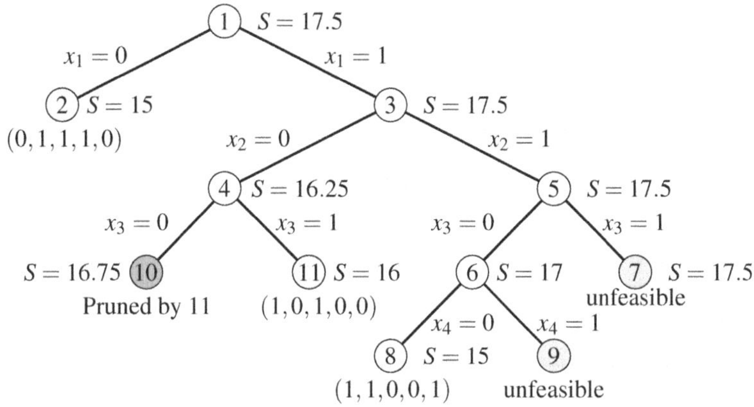

Fig. 4.1 Solving Problem (4.1) with a branch and bound. Sub-problem set Q managed as a stack. The nodes are numbered by creation order. Branching is done by increased variable index. Nodes 9 and 7 are pruned because they cannot lead to feasible solutions. Node 10 is pruned because it cannot lead to a solution better than node 11

solution of the relaxed problem. Frequently, the policy adopted for branching is empirical.

A simple implementation of this framework is the $A*$ *search* algorithm. The last manages Q as a priority queue and evaluates a heuristic value before inserting a sub-problem in Q.

In some cases, the number of possible values for the next x_k variable to set is significant, especially when x_k can take any integer value. A branching technique is to consider the fractional value y taken by a variable x_k and to develop two branches, one with the additional constraint $x_k \leqslant \lfloor y \rfloor$ and the other with $x_k \geqslant \lfloor y \rfloor + 1$. In this case, the sets of fixed and independent variables are unchanged on Line 16. This technique was proposed by Dakin [2].

Another technique, known as *branch and cut* is to add constraints to the relaxed sub-problem. The goal of the new constraints is to remove the unfeasible solution obtained by solving the sub-problem. For instance, such a constraint may prevent a variable to take a given fractional value.

4.1.1.1 Example of Implementation of a Branch and Bound

A naive branch and bound implementation manages the sub-problem set as a stack (policy α). This is performed automatically with a recursive procedure.

For the TSP, a solution is a permutation p of the n cities. The element p_i provides the ith city visited. Assume that the order of cities has been fixed up to and including i and L is the set of cities remaining to be ordered. A lower bound on the optimal tour length can be obtained by considering that:

1. The ith city is connected to the closest of L.

2. Each city of L is connected to another of L that is the closest.
3. The first city is connected to the closest of L.

Doing so, a valid tour is eventually obtained for the complete problem. When only one "free" city remains ($|L| = 1$), we have to go to this one and then to return to the departure city. In this situation, a valid tour is obtained. The procedure given by Code 4.1 returns a flag indicating whether a feasible tour is found.

Code 4.1 tsp_lower_bound.py Code for computing a naive lower bound to the optimal tour. The procedure returns the bound and can alter the order of the last cities of the tour. In the event the length of the modified tour is equal to the value of the lower bound, the procedure indicates that the tour is optimal

```
######### Computation of a naive lower bound for the TSP
def tsp_lower_bound(d,                                  # Distance matrix
                    depth,                 # tour[0] to tour[depth] fixed
                    tour):                                      # TSP tour

    n = len(tour)
    lb = 0 #Compute the length of the path for the cities already fixed in tour
    for j in range(depth):
      lb += d[tour[j]][tour[j+1]]

    valid = 1  # valid is set to 1 if every closest successor of j build a tour
    for j in range(depth, n-1):      # Add the length to the closest free city j
        minimum = d[tour[j]][tour[j+1]]
        for k in range(n-1, depth, -1):
            if k != j and minimum > d[tour[j]][tour[k]]:
                minimum = d[tour[j]][tour[k]]
                if (k > j):
                    tour[k], tour[j+1] = tour[j+1], tour[k]
                else:
                    valid = 0
        lb += minimum

    minimum = d[tour[n-1]][tour[0]]        # Come back to first city of the tour
    for j in range (depth+1, n-1):
        if (minimum > d[tour[j]][tour[0]]):
            valid = 0
            minimum = d[tour[j]][tour[0]]
    lb += minimum
    return lb, tour, valid       # Lower bound, tour modified, lb == tour length
```

To implicitly list all possible tours on n cities, an array as well as a `depth` index can be used. From the `depth` index, all the possible permutations of the last elements of the array are enumerated. This procedure is called recursively with `depth + 1` after trying all the remaining possibilities for the `depth` array entry. To prune the enumeration, no recursive call is performed either if the lower bound computation provides an optimal tour or if the lower bound of the tour length is larger than that of a feasible tour already found. Code 4.2 implements an implicit enumeration for the TSP.

Code 4.2 tsp_branch_and_bound.py Code for implicitly enumerating all the permutations of n elements

```
from tsp_lower_bound import tsp_lower_bound                          # Listing 4.1

######### Basic Branch & Bound for the TSP
def tsp_branch_and_bound(d,                                     # Distance matrix
                         depth,  # current_tour[0] to current_tour[depth] fixed
                         current_tour,                 # Solution partially fixed
                         upper_bound):                      # Optimum tour length

    n = len(current_tour)
    best_tour = current_tour[:]
    for i in range(depth, n):
        tour = current_tour[:]
        tour[depth], tour[i] = tour[i], tour[depth]
        lb, tour, valid = tsp_lower_bound(d, depth, tour)
        if (upper_bound > lb):
            if (valid):
                upper_bound = lb
                best_tour = tour[:]
                print("Improved: ", upper_bound, best_tour)
            else:
              best_tour, upper_bound = tsp_branch_and_bound(d, depth+1, tour, \
                                                            upper_bound)
    return best_tour, upper_bound
```

It should be noted here that this naive approach requires a few seconds to a few minutes to solve problems up to 20 cities. However, this represents a significant improvement over an exhaustive search, which would require a computing time of several millennia. The relaxation based on the notion of 1-tree presented in Sect. 3.1.1.2 could advantageously replace that provided by Code 4.1.

In recent years, so-called exact methods for solving integer linear programs have made substantial progresses. The key improvements are due to more and more sophisticated heuristics for computing relaxations and branching policies. Software like *CPLEX* or *Gurobi* include methods based on metaheuristics for computing bounds or obtaining good solutions. This allows a faster pruning of the enumeration tree. Despite this, the computational time grows exponentially with the problem size.

4.2 Random Construction

A rapid and straightforward method to obtain a solution is to generate it randomly among the set of all feasible solutions. We clearly cannot hope to reliably find an excellent solution like this. However, this method is widely implemented in iterative local searches repeating a constructive phase followed by an improvement phase. It should be noted here that the modeling of the problem plays a significant role, as noted in Chap. 3. In case finding a feasible solution is difficult, one must wonder whether the problem modeling is adequate.

Note that it is not necessarily easy to write a procedure generating each solution of a feasible set with the same probability. Exercise 4.1 deals with the generation of a random permutation of n items. Naive approaches such as those given by Algorithms 4.5 and 4.6 can lead to non-uniform solutions and/or inefficient codes.

4.3 Greedy Construction

In Chap. 2, the first classical algorithms of graphs passed in review—Prim and Kruskal for building the minimum spanning tree and Dijkstra for finding the shortest path—were greedy algorithms. They are building a solution by including an element at every step. The element is permanently added on the base of a function evaluating its relevance for the partial solution under construction.

Assuming a solution is composed of elements $e \in E$ that can be added to a partial solution s, the greedy algorithm decides which element to add by computing an incremental cost function $c(e, s)$. Algorithm 4.2 provides the framework of a greedy constructive method.

Algorithm 4.2: Framework of a greedy constructive method. Strictly speaking, this is not an algorithm since different implementation options are possible, according to the definition of the set E of the elements constituting the solutions and the incremental cost function

Input: A trivial partial solution s (generally $\varnothing$); set E of elements constituting a solution; incremental cost function $c(s,e)$

Result: Complete solution s

$R \leftarrow E$ // Elements that can be added to s

while $R \neq \varnothing$ **do**

  $\forall e \in R$, compute $c(s,e)$

  Choose e' optimizing $c(s,e')$

  $s \leftarrow s \cup e'$ // Include e' in the partial solution s

  Remove from R the elements that cannot be added any more to s

Algorithms with significantly different performances can be obtained according to the definition of E and $c(s, e)$. Considering the example of the Steiner tree, one could consider E as the set of edges of the problem and the incremental cost function as the weight of each edge. In this case, a partial solution is a forest.

Another modeling could consider E as the Steiner points. The incremental cost function would be to calculate a minimum spanning tree containing all terminal nodes plus e and those already introduced in s.

We now provide some examples of greedy heuristics that have been proposed for a few combinatorial optimization problems.

4.3.1 *Greedy Heuristics for the TSP*

Countless greedy constructive methods have been proposed for the TSP. Here is a choice illustrating the variety of definitions that can be made for the incremental cost function.

4.3.1.1 Greedy on the Edges

The most elementary way to design a greedy algorithm for the TSP is to consider the elements e to add to a partial solution s are the edges. The incremental cost function is merely the edge weight. Initially, we start from a partial solution $s = \emptyset$. The set R consists of the edges that can be added to the solution, without creating a vertex of degree > 2 or a cycle not including all the cities. Figure 4.2 illustrates how this heuristic works on a small instance.

4.3.1.2 Nearest Neighbor

One of the easiest greedy methods to program for the TSP is the nearest neighbor. The elements to insert are the cities rather than the edges. A partial solution s is, therefore, a path in which the cities are visited in the order of their insertion. The incremental cost is the weight of the edge that connects the next city. Figure 4.3 illustrates the execution of this heuristic on the same instance as above. It is a coincidence to get a solution identical to the previous method.

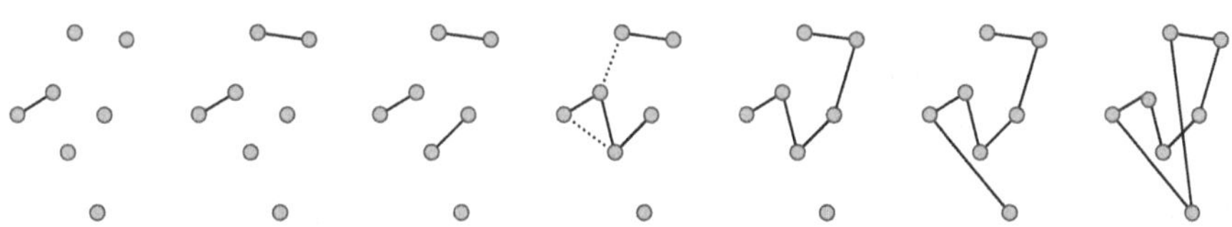

Fig. 4.2 Steps of a greedy constructive method based on the edge weight for the TSP

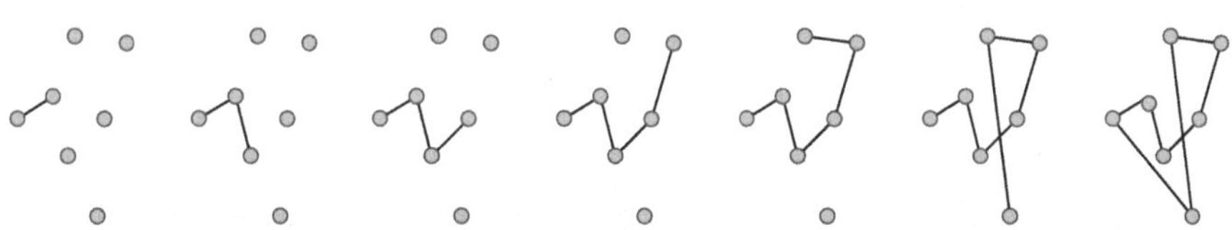

Fig. 4.3 Running the nearest neighbor for a tiny TSP instance

The nearest neighbor greedy heuristic can be programmed very concisely, in $\Theta(n^2)$, where n is the number of cities (see Code 4.3).

Code 4.3 tsp_nearest_neighbor.py Nearest neighbor for the TSP. Note the similarities with the implementation of the Dijkstra algorithm given by Code 2.1

```
######### Nearest Neighbor greedy heuristic for the TSP
def tsp_nearest_neighbor(d,                                       # Distance matrix
                         tour):                                 # List of cities to order

    n = len(tour)
    length = 0                                                        # Tour length
    for i in range(1, n):              # Cities from tour[0] to tour[i-1] are fixed
        nearest = i                                       # Next nearest city to insert
        cost_ins = d[tour[i-1]][tour[i]]                         # City insertion cost
        for j in range(i+1, n):                             # Find next city to insert
            if d[tour[i-1]][tour[j]] < cost_ins:
                cost_ins = d[tour[i-1]][tour[j]]
                nearest = j
        length += cost_ins
        tour[i], tour[nearest] = tour[nearest], tour[i]  # Definitive insertion

    length += d[tour[n - 1]][tour[0]]                              # Come back to start

    return tour, length
```

4.3.1.3 Largest Regret

A defect of the nearest neighbor is to temporarily forget a few cities, which subsequently causes significant detours. This is exemplified in Fig. 4.3. To try to prevent this kind of situation, we can evaluate the increased cost for not visiting city e just after the last city i of the partial path s. In any case, the city e must appear in the final tour. This will cost at least $\min_{j,k\in R} d_{je} + d_{ek}$. Conversely, if e is visited just after i, the cost is at least $\min_{r\in R} d_{ie} + d_{er}$. The largest regret greedy constructive method chooses the city e maximizing $c(s, e) = \min_{j,k\in R} (d_{je} + d_{ek}) - \min_{r\in R} (d_{ie} + d_{er})$.

4.3.1.4 Cheapest Insertion

The cheapest insertion heuristic involves inserting a city in a partial tour. The set E consists of cities, and the trivial initial tour is a cycle on both cities which are the nearest. The incremental cost $c(s, e)$ of a city is the minimum detour that must be consented to insert the city e in the partial tour s between two successive cities of s. Figure 4.4 illustrates this greedy method.

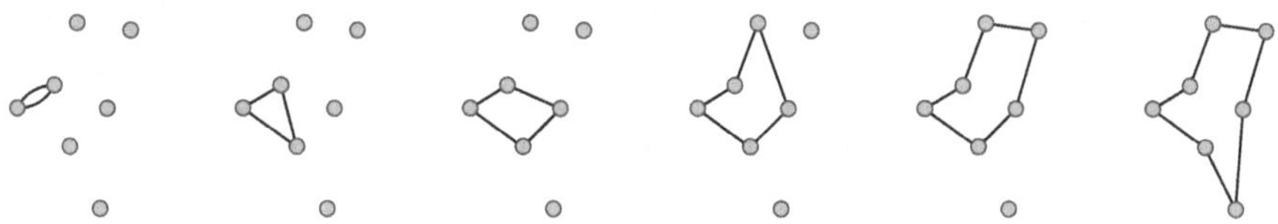

Fig. 4.4 Running the cheapest insertion for a tiny TSP instance

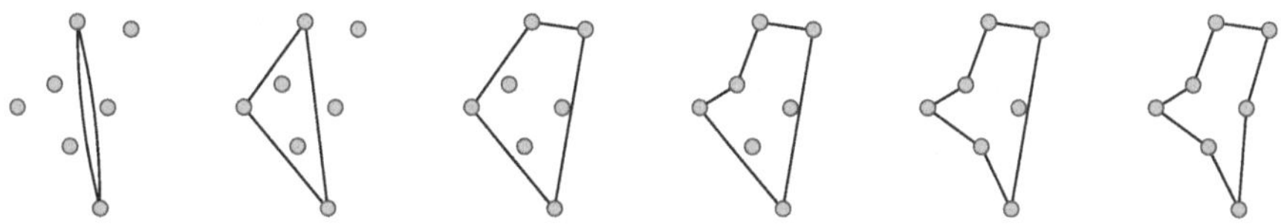

Fig. 4.5 Running the farthest insertion for a tiny TSP instance

4.3.1.5 Farthest Insertion

The farthest insertion heuristic is similar to the previous one, but it selects the city whose insertion causes the most significant detour. However, each city is inserted at the best possible place in the tour. Figure 4.5 illustrates this greedy method. It seems counter-intuitive to choose the most problematic city at each step. However, this type of construction reveals less myopic and frequently produces better final solutions than the previous heuristics.

Here, we have provided only a limited range of greedy constructive methods that have been proposed for the TSP. The quality of the solutions they produce varies. It is usually not challenging to find problem instances for which a greedy heuristic is misguided and makes choices increasingly bad. On points uniformly distributed on the Euclidean plane, they typically provide solutions a few tens of percent above the optimum.

4.3.2 Greedy Heuristic for Graph Coloring

After reviewing several methods for the TSP, it is necessary to present a not too naive example for another problem.

A relatively elaborate greedy method for coloring the vertices of a graph tries to determine the node for which assigning a color may be the most problematic. The DSatur [1] method assumes it corresponds to the node with already colored neighbors using the broadest color palette. For this purpose, the *saturation* degree of a vertex v is defined, noted DS (v), corresponding to the number of different colors used by the vertices adjacent to v. At equal degree of saturation—particularly at the start, when no vertex is colored—the node with the highest degree is selected. At equivalent degree and saturation degree, the nodes are arbitrarily selected. Algorithm 4.3 formalizes this greedy method.

Algorithm 4.3: *DSatur* algorithm for graph coloring. The greedy criterion used by this algorithm is the saturation degree of the vertices, corresponding to the number of different colors used by adjacent nodes

Input: Undirected graph $G = (V, E)$;
Result: Vertex coloring
Color with 1 the vertex v with the highest degree
$R \leftarrow V \setminus v$
$colors \leftarrow 1$
while $R \neq \varnothing$ **do**
    $\forall v \in R$, compute $DS(v)$
    Choose v' maximizing $DS(v')$, with the highest possible degree
    Find the smallest k $(1 \leqslant k \leqslant colors + 1)$ such that color k is feasible for v'
    Assign color k to v'
    if $k > colors$ **then**
        $colors = k$
    $R \leftarrow R \setminus v'$

4.4 Improvement of Greedy Procedures

The chief drawback of a greedy construction is that it never changes a choice performed in a myopic way. Conversely, the shortcoming of a complete enumerative method is the exponential growth of the computational effort with the problem size. To limit this growth, it is therefore necessary to limit the branching. This is typically achieved on the basis of greedy criteria. This section reviews two partial enumeration techniques that have been proposed to improve a greedy algorithm.

First, the beam search was proposed within the framework of an application in speech recognition [5]. Second is the more recent pilot method, proposed by Duin and Voß [3]. It was presented as a new metaheuristic. Other frames have been derived from it [4].

4.4.1 Beam Search

Beam search is a partial breadth-first search. Instead of keeping all the branches, at most, p are kept at each level, on the basis of the incremental cost function $c(s, e)$. Arriving at level k, the partial solution at the first level is completed with the element e which leads to the best solution at the last enumerated level. Figure 4.6 illustrates the principle of a beam search.

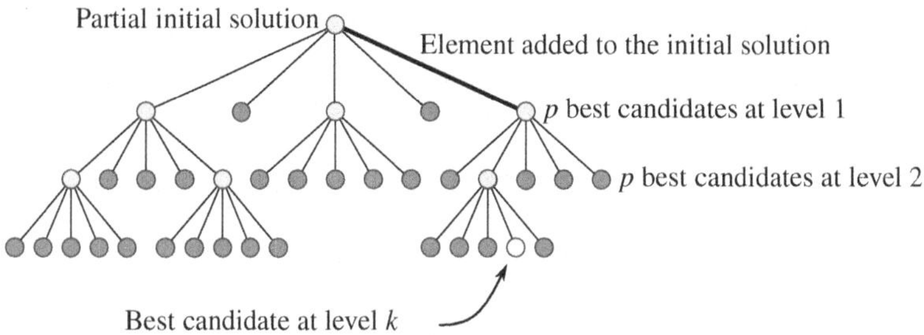

Fig. 4.6 Beam search with $p = 3$ and $k = 3$. Before definitively choosing the element to insert in the partial solution, a breadth-first search is carried out up to a depth of k, only retaining the p best candidates at each depth

A beam search variant proceeds by making a complete enumeration up to a level containing more than p nodes. The p best of them are retained to generate the candidates for the next level.

4.4.2 *Pilot Method*

The framework of the pilot method requires a so-called pilot heuristic to fully complete a partial solution. This pilot heuristic can be a simple greedy method, for example, the nearest neighbor heuristic for the TSP, but it can equally be a much more sophisticated method, such as one of those presented in the following chapters.

The pilot method enumerates all the partial solutions that can be obtained by including an element to the starting solution. The pilot heuristic is then applied to all these partial solutions to end up with as many complete solutions. The partial solution at the origin of the best complete solution is used as the new starting solution, until there is nothing more to add. Figure 4.7 illustrates two steps of the method.

Algorithm 4.4 specifies how the pilot metaheuristic works. In this framework, the ultimate "partial" solutions represent a feasible complete solution which is not necessarily the solution returned by the algorithm. Indeed, the pilot heuristic can generate a complete solution that does not necessarily include the elements of the initial partial solution, especially if it includes an improvement technique more elaborated than a simple greedy constructive method.

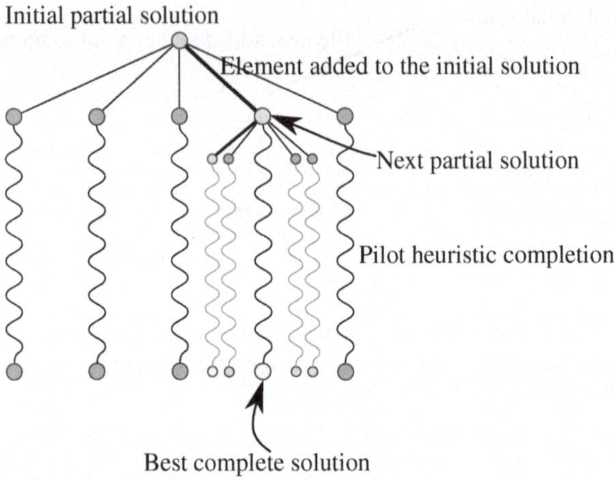

Fig. 4.7 Pilot method. An element is included in the partial solution; then a pilot constructive heuristic is applied to fully complete it. The process is repeated with another element added to the partial solution. The element finally inserted is the one that led to the best complete solution

---

Algorithm 4.4: Frame of a pilot method

Input: s_p trivial partial solution; set E of elements constituting a solution; pilot heuristic $h(s_e)$ for completing a partial solution s_e; fitness function $f(s)$
Result: Complete solution $s*$

```
R ← E                                                  // Elements that can be added to s
while R ≠ ∅ do
    v ← ∞
    forall e ∈ R do
        Complete s_p with e to get s_e
        Apply h(s_e) to get a complete solution s
        if f(s) ⩽ v then
            v ← f(s)
            s_r ← s_e
            if s is better than s* then Store the improved solution
                s* ← s
    s_p ← s_r                                          // Add an element to the partial solution s_p
    Remove from R the elements that cannot properly be added to s_p
```

Code 4.4 provides an implementation of the pilot method for the TSP. The pilot heuristic is the nearest neighbor.

Code 4.4 tsp_pilot_nearest_neighbor.py Implementation of a pilot method with the nearest neighbor (Code 4.3) as pilot heuristic

```
from tsp_utilities import tsp_length                              # Listing 12.2
from tsp_nearest_neighbor import *                                # Listing  4.3

######### Constructive algorithm with Nearest Neighbor as Pilot method
def tsp_pilot_nearest_neighbor(n,                              # Number of cities
                               d):                             # Distance matrix
    tour = [i for i in range(n)]                     # All cities must be in tour

    for q in range(n - 1):             # Cities up to q at their final position
        length_r = tsp_length(d, tour)
        to_insert = q
        for r in range(q, n):        # Choose next city to insert at position q
            sol = [tour[i] for i in range(n)]                  # Copy of tour in sol
            sol[q], sol[r] = sol[r], sol[q]         # Tentative city at postion q
            sol[q:n], _ = tsp_nearest_neighbor(d, sol[q:n])
            tentative_length = tsp_length(d, sol)
            if length_r > tentative_length:
                length_r = tentative_length
                to_insert = r

        # Put definitively to_insert at position q
        tour[q], tour[to_insert] = tour[to_insert], tour[q]

    return tour, tsp_length(d, tour)
```

Problems

4.1 Random Permutation

Write a procedure to generate a random permutation of n elements contained in an array p. It is desired a probability of $1/n$ to find any element in any position in p. Describe the inadequacy of Algorithms 4.5 and 4.6.

Algorithm 4.5: Bad algorithm to generate a random permutation of n elements

Input: A set of n elements $e_1, \dots, e_n$
Result: A permutation p of the elements

1 $i \leftarrow 0$ // Number of element already chosen
2 **while** $i \neq n$ **do**
3   Draw a random number u uniformly between 1 and n
4   **if** e_u *is not already chosen* **then**
5     $i \leftarrow i + 1$
6     $p_i \leftarrow e_u$

Algorithm 4.6: Another bad algorithm to generate a random permutation of n elements

Input: A set of n elements $e_1, \ldots, e_n$
Result: A permutation p of the elements

$i \leftarrow 0$ // Number of element already chosen
while $i \neq n$ **do**
  Draw a random number u uniformly between 1 and n
  $i \leftarrow i + 1$
  if e_u *is already chosen* **then**
    Find the next u' such that $e_{u'}$ is not chosen
    $p_i \leftarrow e_{u'}$
  else
    $p_i \leftarrow e_u$

4.2 Greedy Algorithms for the Knapsack
Propose three different greedy algorithms for the knapsack problem.

4.3 Greedy Algorithm for the TSP on the Delaunay
We want to build the tour of a traveling salesman (on the Euclidean plane) using only edges belonging to the Delaunay triangulation. Is this always possible? If this is not the case, provide a counter-example; otherwise, propose a greedy method and analyze its complexity.

4.4 TSP with Edge Subset
To speed up a greedy method for the TSP, only the 40 shortest edges adjacent to each vertex are considered. Is this likely to reduce the algorithmic complexity of the method? Can this cause some issues?

4.5 Constructive Method Complexity
What is the complexity of the nearest neighbor heuristic for TSP? Same question if we use this heuristic in a beam search by retaining p nodes at each depth and that we go to k levels down. Similar question for the pilot method where we equally employ the nearest neighbor as the pilot heuristic.

4.6 Beam Search and Pilot Method Applications
We consider a TSP instance on five cities. Table 4.1 gives its distance matrix.

Apply a beam search to this instance, starting from the city 1. At each level, only $p = 2$ nodes are retained, and the tree is developed up to $k = 3$ levels down.

Apply a pilot method to this instance, considering the nearest neighbor as the pilot heuristic.

Table 4.1 Distance matrix for Problem 4.6

	1	2	3	4	5
1	—	5	3	19	7
2	13	—	1	18	6
3	12	4	—	14	6
4	11	9	8	—	10
5	23	11	7	21	—

4.7 Greedy Algorithm Implementation for Scheduling
Propose two greedy heuristics for the permutation flowshop problem. Compare their quality on problem instances from the literature.

4.8 Greedy Methods for the VRP
Propose two greedy heuristic methods for the vehicle routing problem.

References

1. Brélaz, D.: New methods to color the vertices of a graph. Commun. ACM. **22**(4), 251–256 (1979). https://doi.org/10.1145/359094.359101
2. Dakin, R.J.: A tree search algorithm for mixed integer programming problems. Comput. J. **8**(3), 250–255 (1965). https://doi.org/10.1093/comjnl/8.3.250
3. Duin, C., Voß S.: The pilot method: a strategy for heuristic repetition with application to the steiner problem in graphs. Networks. **34**, 181–191 (1999). https://doi.org/10.1002/(SICI)1097-0037(199910)34:3%3C181::AID-NET2%3E3.0.CO;2-Y
4. Furcy, D., Koenig, S.: Limited discrepancy beam search. In: Proceedings of the International Joint Conference on Artificial Intelligence (IJCAI), pp. 125–131 (2005)
5. Lowerre, B.: The Harpy Speech Recognition System. Ph. D. Thesis, Carnegie Mellon University (1976)

Chapter 5
Local Search

Improvement techniques are a cornerstone in the design of heuristics. As will be seen later, most metaheuristics incorporate a local search.

By examining the solutions produced by greedy constructive heuristics like those presented in the previous chapter, we immediately notice they are not optimal. For example, the solution of the Euclidean traveling salesman problem obtained by the nearest neighbor heuristic, shown in Fig. 4.3, has intersecting edges, which is obviously suboptimal. Indeed, it is possible to replace the two intersecting edges by two others whose sum of the lengths is lower while preserving a tour. Replacing two edges by two others that are shorter can then be repeated until a solution cannot be improved by the same process as shown in Fig. 5.1.

5.1 Local Search Framework

The general idea is therefore to start from a solution obtained using a constructive method and improve it locally. The process is repeated until no further improvement is achieved. This frame is well known in continuous optimization, to seek an optimum of a differentiable function with gradient methods. The gradient concept for finding an improving direction does not exist in discrete optimization; it is replaced by the definition of "minor" changes of the solution called *moves* or the concept of *neighborhood*.

To be able to apply Algorithm 5.1, it is necessary to define how to obtain the neighbor solutions. Formally, a neighborhood set $N(s) \subset S$ must be defined for any solution $s \in S$. Therefore, the search for a modification of s implies enumerating the solutions of $N(s)$ to extract one of them, s', which is better than s.

A convenient way to define a neighborhood $N(s)$ is to specify the modifications, commonly called the moves, that can be applied to the solution s. In the example of Fig. 5.1 for the TSP, a move m can be specified by a pair of cities $[i, j]$. It consists

É. D. Taillard, *Design of Heuristic Algorithms for Hard Optimization*, Graduate Texts in Operations Research, https://doi.org/10.1007/978-3-031-13714-3_5

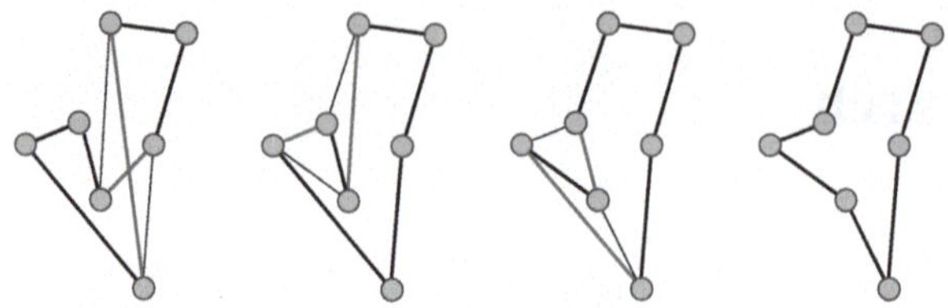

Fig. 5.1 Successive improvements of a TSP solution with the 2-opt local search. Two edges (in color) are replaced by two others (dotted), and the sum of their lengths is lower

---

Algorithm 5.1: General framework of a local improvement method. It is assumed to have a solution to the problem as well as a method able to generate, from any solution, a number of other ones

---

Input: Solution s, method modifying a solution
Result: Improved solution s

repeat
  if *there is a modification of* s *into* s' *improving* s **then**
    $s \leftarrow s'$
until *no improvement of* s *is found*

---

in replacing the edges $[i, s_i]$ and $[j, s_j]$ by the edges $[i, j]$ and $[s_i, s_j]$, where s_i and s_j are, respectively, the cities that follow i and j in the solution s.

This neighborhood of replacing two edges by two others is known in the literature as *2-exchange* or *2-opt* [2]. The set $M(s)$ of 2-opt moves that can be applied to the solution s can be formally defined by $M(s) = \{[i, j], i, j \in s, i \neq j, j \neq s_i, i \neq s_j\}$. Applying a move $m \in M(s)$ to the solution s is sometimes noted $s \oplus m$.

The definition of the neighborhood can be obtained with the definition of the set of moves: $N(s) = \{s' | s' = s \oplus m, m \in M(s)\}$. The size of the 2-opt neighborhood is $|N(s)| = \Theta(|s|^2)$. The application of an improvement method to the TSP can therefore be reasonably achieved by enumerating the neighbor solutions. This enumeration can be done according to two policies, either the *first improvement* or the *best improvement*.

5.1.1 First Improvement Heuristic

With this policy, the current solution is immediately changed as soon as an improving move is identified. The neighborhood is therefore not thoroughly examined at each iteration. This policy is therefore aggressive. It allows a solution to be improved quickly. It generally leads to a greater number of changes to the initial solution than the best improvement policy. The framework of the first improvement method is provided by Algorithm 5.2.

Algorithm 5.2: Framework of the first improvement heuristic

```
Input: Solution s, neighborhood specification N(·), fitness function f(·) to minimize.
Result: Improved solution s
forall s′ ∈ N(s) do
    if f(s′) < f(s) then Move to s′, break the loop and initiate the next one
        s ← s′
```

5.1.2 Best Improvement Heuristic

With the best improvement policy, the neighborhood is thoroughly examined at each iteration. The best neighbor solution identified is the current one for the subsequent iteration. Algorithm 5.3 formalizes this policy.

Algorithm 5.3: Framework of the best improvement method

```
Input: Solution s, neighborhood specification N(·), fitness function f(·) to minimize.
Result: Improved solution s
repeat
    end ← true
    best_neighbor_value ← ∞
    forall s′ ∈ N(s) do
        if f(s′) < best_neighbor_value then A better neighbor is found
            best_neighbor_value ← f(s′)
            best_neighbor ← s′
    if best_neighbor_value < f(s) then Move to the improved solution
        s ← best_neighbor
        end ← false
until end
```

It performs more work between each change to the solution. The improvements are therefore larger and fewer in number. This policy is less frequently used in metaheuristics. We will see later that taboo search is based on this framework. Indeed, this technique tries to learn how to modify a solution smartly. It is therefore appropriate to examine the neighborhood thoroughly rather than rushing to the first small improvement encountered.

However, there are problems where there is an interest in exploiting this policy. For the quadratic assignment problem, it can be shown that evaluating a neighbor solution takes a time proportional to $O(n)$, whereas it is possible to evaluate the set of $O(n^2)$ neighbor solutions in $O(n^2)$. With the same computational effort, it is therefore possible to evaluate more solutions with the best improvement policy.

Sometimes, the set $N(s)$ is so large that its enumeration is done either implicitly to extract the best neighbor solution or heuristically; we will come back to this in particular in Chap. 6 with the large neighborhood search technique Sect. 6.4.1.

Code 5.1 implements the best improvement framework for the TSP. The algorithm seeks the best replacement of two edges by two others.

Code 5.1 tsp_2opt_best.py Implementation of a best improvement method for the TSP with 2-opt neighborhood

```
######### Local search with 2-opt neighborhood and best improvement policy
def tsp_2opt_best(d,                                          # Distance matrix
                  tour,                                              # Solution
                  length):                                    # Solution length
    n = len(tour)
    best_delta = -1
    while best_delta < 0:
        best_delta = float('inf')
        best_i = best_j = -1                                # Best move to perform
        for i in range(n - 2):
            j = i + 2
            while j < n and (i > 0 or j < n - 1):
                delta = \
                    d[tour[i]][tour[j]]    + d[tour[i+1]][tour[(j+1)%n]] \
                  - d[tour[i]][tour[i+1]] - d[tour[j]][tour[(j+1)%n]]
                if delta < best_delta:
                    best_delta = delta
                    best_i, best_j = i, j
                j += 1                                         # Next neighbor

        if best_delta < 0:      # Perform best move if it improves best solution
            length += best_delta                        # Update solution cost
            i, j = best_i+1, best_j      # Reverse path from best_i+1 to best_j
            while i < j:
                tour[i], tour[j] = tour[j], tour[i]
                i, j = i + 1, j - 1

    return tour, length
```

5.1.3 *Local Optima*

By applying Algorithm 5.1, whether one of its variants (5.2 or 5.3), a *locally optimal* solution is obtained with respect to the neighborhood used. Indeed, there is no guarantee that the returned solution is the best for the particular instance. *Globally optimal* solutions are therefore opposed to those which are only locally optimal. It should be emphasized that a local optimum relative to one neighborhood is not necessarily a local optimum for another neighborhood (see Problem 5.1). A *plateau* is a set of neighboring solutions all having the same value. Figure 5.2 illustrates the notion of local optimum, plateau, and global optimum.

Objective functions such as min(max(...)) generate many plateaus with many solutions. Their optimization with a local search is therefore difficult. Indeed, all solutions in a plateau are local optima.

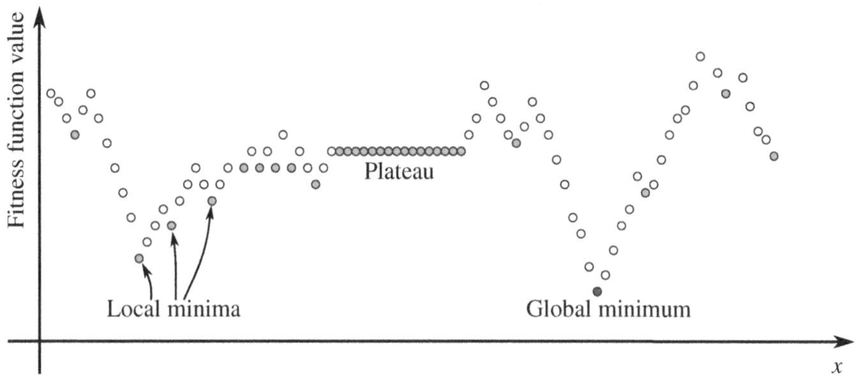

Fig. 5.2 Local minima, global minimum, and plateau of a function of a discrete variable x relative to a neighborhood consisting in changing x by one unit

However, for some problems, that frame provides globally optimal solutions. These include, in particular, the shortest path problem with the Bellman-Ford Algorithm 2.4 and linear programming with the simplex algorithm.

Since the set of solutions to a combinatorial problem is finite and Algorithms 5.2 and 5.3 only modify the solution if it is strictly improved, we deduce these algorithms end after a finite time. By cons, their calculation time is not necessarily polynomial, even if the size of the neighborhood is. In practice, as with the simplex algorithm, we do not observe such a degeneration.

Figure 5.3 shows the evolution of the average computing time for two methods applied to the TSP (the best improvement and first improvement policies) as a function of the number of cities. The starting solution is randomly generated, and the instances are selected from the 2D Euclidean ones of the TSPLIB [11].

5.1.3.1 TSP 3-Opt

The beginning of this chapter presents a local search for the TSP based on replacing two edges by two others. It is naturally possible to define other neighborhoods, for instance, replacing three edges (or arcs in the case of a non-symmetrical problem) by three others. Figure 5.4 shows this type of move, called 3-opt.

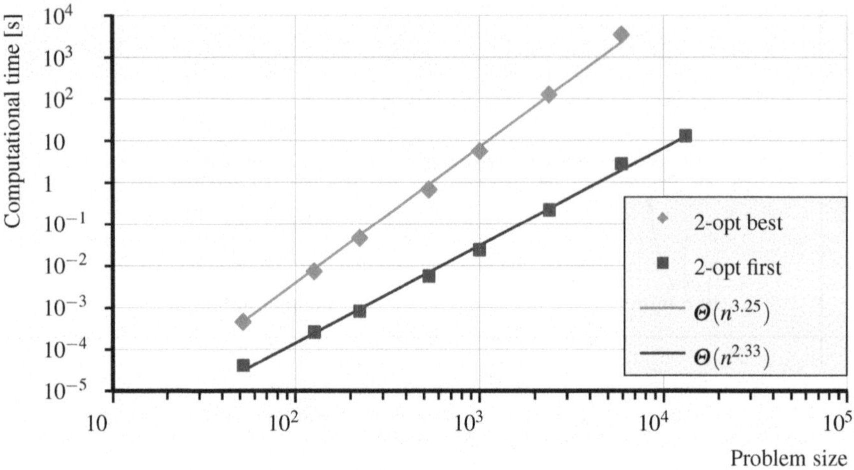

Fig. 5.3 Evolution of the computational time as a function of the number of TSP cities for two variants of improvement methods. Both scales are logarithmic. Hence, computational times approximately aligned on a straight line indicates a polynomial dependence

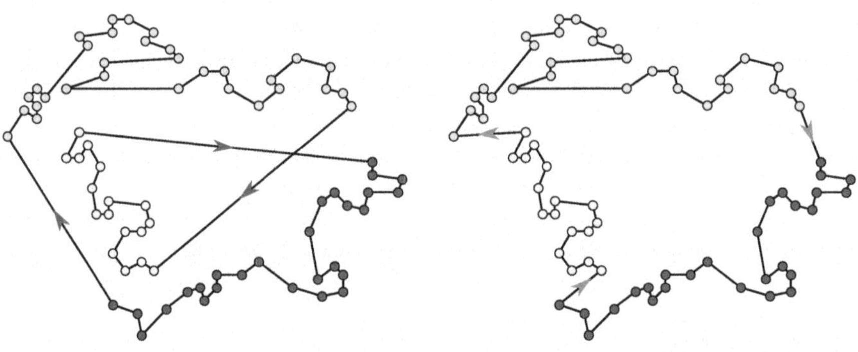

Fig. 5.4 3-opt move. Three arcs are replaced by three others. They connect three sub-paths traversed in the same direction, before and after modification. Another way to apprehend the 3-opt neighborhood is the displacement of a sub-path elsewhere in the tour

Code 5.2 tsp_3opt_first.py Implementation of an improvement method based on 3-opt neighborhood. In practice, its complexity in $\Omega(N^3)$ is excessive for tackling instances with several hundred cities

```
######### Local search with 3-opt neighborhood and first improvement policy
def tsp_3opt_first(d,                                    # Distance matrix
                   succ,                                        # Solution
                   length):                              # Solution length

    last_i, last_j, last_k = 0, succ[0], succ[succ[0]]
    i, j, k = last_i, last_j, last_k
    while True:
        delta = d[i][succ[j]] + d[j][succ[k]] + d[k][succ[i]] \
               -d[i][succ[i]] - d[j][succ[j]] - d[k][succ[k]]      # Move cost
        if delta < 0:                               # is there an improvement?
            length += delta                             # Update solution cost
            succ[i], succ[j], succ[k] = succ[j], succ[k], succ[i]# Perform move
            j, k = k, j                             # Replace j between i and k
            last_i, last_j, last_k = i, j, k
        k = succ[k]                                                 # Next k
        if k == i:                             # k at its last value, next j
            j = succ[j]; k = succ[j]
        if k == i:                             # j at its last value, next i
            i = succ[i]; j = succ[i]; k = succ[j]
        if i == last_i and j == last_j and k == last_k:
          break

    return succ, length
```

An attractive property of this neighborhood is not to change the path direction between the three nodes whose successors are modified. In the case of symmetrical problems, there are several ways to reconnect the three sub-paths (see Problem 5.4).

Representing a solution by means of a permutation s whose element s_i indicates the city visited just after the city i, a 3-opt move can be implemented in constant time. Checking that a solution is 3-optimal can be done in $O(n^3)$. Hence, without using neighborhood reduction techniques, it can only manage relatively small instances. Code 5.2 implements a local search for the TSP with 3-opt moves. It applies the first improvement policy.

5.1.3.2 TSP Or-Opt

Another type of neighborhood proposed by Or [8] is to modify the visiting order of a few consecutive cities. The idea is to examine whether it is pertinent to place three successive cities somewhere else in the current tour.

The originality of the method proposed by Or is exploiting several neighborhoods. Once it is no longer possible to improve the solution by changing the visiting order of three cities, we try to change only two. As soon as changing two cities improves the solution, we return to the first neighborhood. When the solution is locally optimal with respect to these two neighborhoods, we try changing the position of a unique city. Figure 5.5 illustrates a possible Or-opt move.

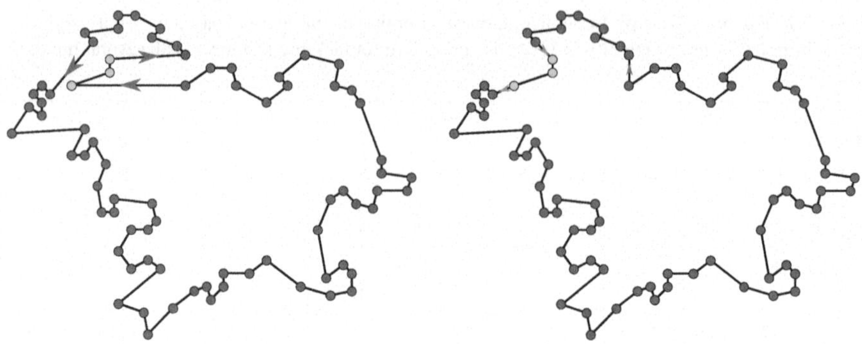

Fig. 5.5 Or-opt move, where three successive vertices are moved within the tour. Possibly, it is worthier to modify the sub-path direction

This neighborhood is distinct from a limitation of the 3-opt neighborhood in which a sub-road would be limited to 3, 2, or 1 city. Indeed, Or neighborhood tries to reverse the sub-path direction. Testing if a tour is Or-optimal takes $\Theta(n^2)$ time.

5.1.3.3 Data Structure for TSP 2-Opt

The 2-opt neighborhood reverses the direction of a sub-path. Visually, this seems innocuous for a symmetrical instance. But, it is not so for the computer representation of a solution. A data structure, inspired by the work of [9], enables performing a 2-opt move in constant time. For each city, an array stores both adjacent cities. The array components at indices $2i$ and $2i + 1$ provide both adjacent cities of the city i.

An array t with $2n$ indices represents a solution. Initially, $t_{2i}/2$ provides the number of the city succeeding i and $(t_{2i+1} - 1)/2$ the number of the city preceding i. A 2-opt move consists in modifying four values of the array t. This can be realized in constant time. Figure 5.6 illustrates the operating principle of this data structure.

Code 5.3 initializes an array t implementing this data structure from a given tour. The last is provided by the list of cities successively visited (and not the successor of each city).

Code 5.4 implements a first improvement heuristic based on this principle. It uses the shift operator `i >> 1` to quickly evaluate the expression `i/2`. The last is the number of the ith city. The "exclusive or" operator `i ^ 1` allows quickly calculating the expression `i+1-2*(i%2)` providing the index to access the adjacent city.

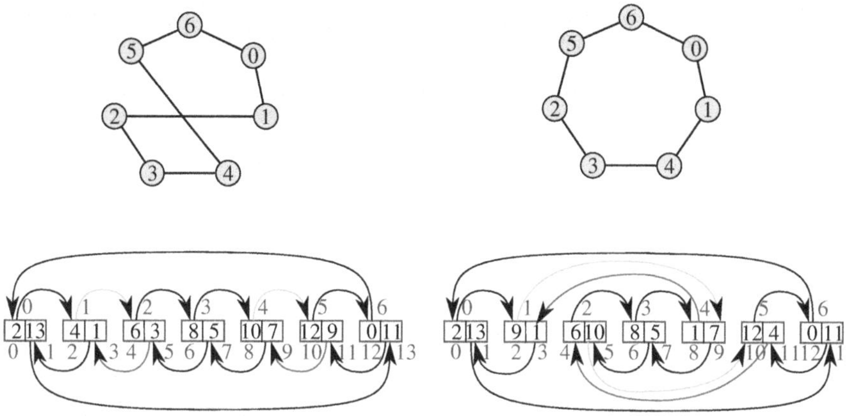

Fig. 5.6 Data structure for performing 2-opt moves in constant time. The $2i$ and $2i + 1$ entries of an array allow identifying both cities adjacent to the city i. Starting from the index 0, we can reconstitute the tour by following the adjacent city. Starting from the index 1, we can reconstitute the tour in the other direction. Performing a 2-opt move consists in altering four entries in the array

Code 5.3 build_2opt_data_structure.py Implementation and initialization of the data structure presented in Fig. 5.6. The tour is provided by the sequence of the cities to visit. The data structure allows performing a 2-opt move in a constant time

```
######### Data structure building for performing 2-opt move in constant time
def build_2opt_data_structure(tour):                    # Order of visit of the cities
    n = len(tour)
    t = [-1] * 2 * n
    for i in range(n - 1):                                              # Forward tour
        t[2 * tour[i]] = 2 * tour[i + 1]
    t[2 * tour[n - 1]] = 2 * tour[0]
    for i in range(1, n):                                              # Backward tour
        t[2 * tour[i] + 1] = 2 * tour[i - 1] + 1
    t[2 * tour[0] + 1] = 2 * tour[n - 1] + 1
    return t
```

Code 5.4 tsp_2opt_first.py Implementation of a first improvement heuristic for the TSP based on 2-opt neighborhood. This implementation exploits the data structure shown in Fig. 5.6 for performing the moves in constant time

```
from build_2opt_data_structure import build_2opt_data_structure  # Listing 5.3
from tsp_utilities import tsp_2opt_data_structure_to_tour         # Listing 12.2

######### Local search with 2-opt neighborhood and first improvement policy
def tsp_2opt_first(d,                                          # Distance matrix
                   tour,                                              # Solution
                   length):                                    # Solution length

    n = len(tour)
    t = build_2opt_data_structure(tour)
    i = last_i = 0              # i = starting city || last_i = i - a complete tour
    while t[t[i]] >> 1 != last_i:  # Index i has made 1 turn without impovement
        j = t[t[i]]
        while j >> 1 != last_i and (t[j]>>1 != last_i or i>>1 != last_i):
            delta =  d[i>>1][j>>1]     + d[t[i]>>1][t[j]>>1] \
                   - d[i>>1][t[i]>>1] - d[j>>1][t[j]>>1]
            if delta < 0:                             # An improving move is found
                next_i, next_j = t[i], t[j]                         # Perform move
                t[i], t[j] = j ^ 1, i ^ 1
                t[next_i ^ 1], t[next_j ^ 1] = next_j, next_i

                length += delta                           # Update solution cost
                last_i = i >> 1    # Solution improved: i must make another turn
                j = t[i]
            j = t[j]                                                    # Next j
        i = t[i]                                                        # Next i

    return tsp_2opt_data_structure_to_tour(t), length
```

5.1.4 Neighborhood Properties

A neighborhood connects various solutions of the problem. Thus, a graph can represent it. The vertices are the solutions. An edge connects two neighbor solutions. The edges can be directed if the moves are not immediately reversible. An efficient neighborhood, or its representative graph, should have certain properties.

5.1.4.1 Connectivity

The connectivity property of a neighborhood stipulates that any feasible solution can reach at least one globally optimal solution. In other words, there must be a path from any vertex to one representing an optimal solution. This path is not necessarily monotonously improving.

5.1.4.2 Low Diameter

A neighborhood is expected to allow discovering an optimal solution in a few steps. By definition, the diameter of a graph is the maximum length of a shortest path connecting two vertices.

5.1.4.3 Low Ruggedness

A neighborhood should have as few local optima as possible and a strong correlation between the values of neighbor solutions. The ideal would be to have only one, which would then be the global optimum, achievable every time starting from any solution. This property is certainly not satisfied for intractable problems and polynomial neighborhoods. However, finding an adequate neighborhood for the problem under consideration is essential for the success of an improvement method.

For problems like the TSP, many neighborhoods have been devised, some being remarkably effective for obtaining excellent solutions. This can most likely be explained by the visual side of the problem, which considerably supports us in deciding what modifications to make to a solution to improve it. For other problems, it is challenging to imagine neighborhoods, and these sometimes lead to "egg carton"-type landscapes, very poorly suited to optimization.

One possibility for smoothing a neighborhood is to modify the fitness function. The *flying elephant* method of [13] exploits this trick for the optimization of continuous functions. The $|x|$ terms in the objective function are replaced by $\sqrt{x^2 + \tau^2}$ and terms of the type $max(0, x)$ by $(x + \sqrt{x^2 + \tau^2})/2$. When the parameter τ tends toward 0, the modified fitness function gets closer to the original objective function. Flying elephants is a metaphor that makes the analogy with a large round object dropped on a rugged field. Such an object will roll further to a lower altitude than a small one that will stop at the first small basin.

5.1.4.4 Small Size

Since improvement methods are based on the systematic evaluation of neighbor solutions, the neighborhood size should not be excessively large. For instance, the 2-opt neighborhood for the TSP is in $O(n^2)$. This allows addressing problems with several thousand cities. This is not possible with the 3-opt neighborhood, in $O(n^3)$.

5.1.4.5 Fast Evaluation

Finally, the algorithmic complexity of evaluating the quality of the neighbor solutions should be as low as possible. For an asymmetric traveling salesman problem, the advantage of the small 2-opt neighborhood size is negated by the fact that a constant time cost evaluation is no longer possible. In this case, evaluating

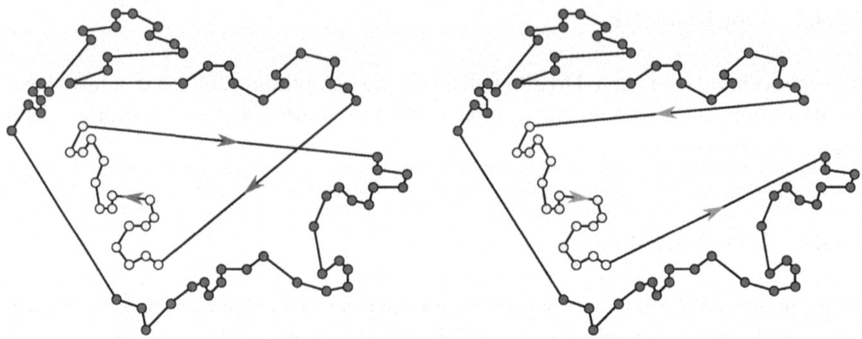

Fig. 5.7 2-opt move where two edges are replaced by two other edges. This type of move is only suitable for symmetrical problems because a part of the tour is traversed in the opposite direction, which can significantly alter its cost if the problem is not symmetrical

the larger 3-opt neighborhood could be faster. As shown in Fig. 5.7, part of the tour is reversed with the 2-opt neighborhood. Thus, the ruggedness of the 2-opt neighborhood is also higher than that of the 3-opt for highly asymmetric problems.

5.2 Neighborhood Limitation

Typically, the size of a neighborhood grows quadratically or cubically with the size of the problem instance. As local searches require repeatedly evaluating the entire neighborhood, the computations are prohibitive as the instance size increases. Various techniques have been proposed to limit the computational growth.

5.2.1 Candidate List

A first idea is to make the hypothesis that a favorable move for a solution will remain good for similar solutions. A general method for limiting the computational effort is first to evaluate all moves applicable to a given solution. A selection of the best ones is stored in a candidate list. Only the moves contained in the list are evaluated for some iterations. Periodically, the whole neighborhood must be evaluated. Indeed, the solution is likely to have been quite modified since the candidate list elaboration. Moves that were unfavorable can thus become interesting and vice versa.

5.2.1.1 Candidate List for the Euclidean TSP

In the case of the TSP, the move evaluation is independent of the solution. A candidate list of moves does not need to be periodically reconstructed. But, it implies developing a mechanism to detect whether a given move is valid. For instance, a move can create two or more sub-cycles. If the TSP cities are on the Euclidean plane, building a Delaunay triangulation requires a work in $O(n \log n)$.

The candidate moves only consider the edges present in the triangulation. It can be proved that a Delaunay triangulation has $\Theta(n)$ edges and an average vertex degree not exceeding six. Hence, the size of this limited neighborhood is in $\Theta(n)$. Empirical observation reveals the edges of an optimal tour are almost all part of the Delaunay triangulation. This is illustrated in Fig. 5.8.

Unluckily, this technique solely applies to Euclidean problems. Indeed, the construction of a Delaunay triangulation relies on geometric properties. A general neighborhood limitation technique uses a form of learning with solutions (see Chap. 10) or vocabulary building (see Sect. 8.2).

The idea behind this technique is to generate a number of solutions and limit the neighborhood to moves comprising only elements making part of these solutions. They do not need to be of exceptional quality. But, they must have diverse structures and have similar portions with good solutions to the problem. Also, obtaining them should not need excessive computational effort. Chapter 6 shows how to proceed with large instances.

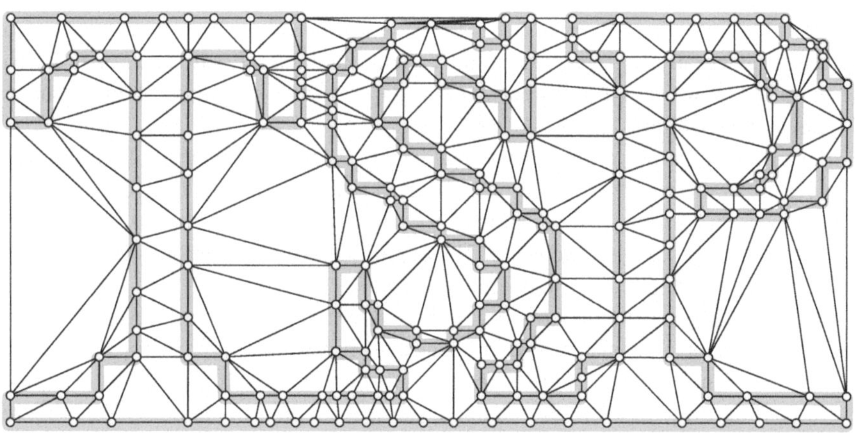

Fig. 5.8 Optimal tour of the TSP instance *tsp225* on which is superimposed the Delaunay triangulation. Here, all the edges of the optimal tour are part of the triangulation

5.2.1.2 TSP Neighborhood Limitation with 1-Trees

Another general neighborhood limitation technique for the TSP, proposed by Helsgaun [5], uses the concept of 1-tree (see Sect. 3.1.1.2 presenting a Lagrangian relaxation for the TSP). Shortly, the idea is to compute the extra cost that would result if the edge (i, j) is part of the 1-tree. The neighborhood reduction proceeds by keeping a few edges—typically 5—adjacent to each vertex with the lowest extra costs. A beneficial collateral effect of this edge length modification technique is to lower the ruggedness of the neighborhood.

Its algorithmic complexity depends on the construction of a minimum spanning tree. As seen in Sect. 2.1, this complexity depends on the number of edges. For large instances, it is necessary to start by reducing their number, for example, with the merging tour technique presented in the previous section. Both neighborhood reduction techniques have similarities with granular search.

5.2.2 *Granular Search*

Granular search is to a priori eliminate solutions with certain characteristics. Illustrating this on the vehicle routing problem, one can assume that good solutions will not include a path directly connecting distant customers. For this problem, [12] proposed to ignore the solutions which involve trips between two clients whose length is greater than a value. The last is set to β times the average trip length of a solution obtained using a fast constructive heuristic, where β is a parameter generally slightly greater than 1. This parameter is called the local search *granularity*. However, the paths between the depot and the customers should remain, whatever their length is.

A similar technique has been used extensively for the TSP. Instead of considering a complete graph where each city is connected to all others, it is only connected to its p closest neighbors, with p limited to a few dozen. Thus, the size of a 2-opt neighborhood is $n \cdot p^2$ instead of n^2. The quality loss of the solutions obtained with such a neighborhood reduction is often negligible.

By cons, implementation of this idea is not trivial. First, the reduced graph where each node is connected to its p nearest neighbors may be not connected. It is therefore necessary to add longer edges so that the graph contains at least one cycle passing through all nodes. Second, the local search implementation is more complex.

For instance, the data structure shown in Sect. 5.1.3.3 for the 2-opt neighborhood cannot be used directly. Indeed, it is fast to determine the city s_i succeeding to city i and a city j close to i (there are only p candidates). But it is not possible to immediately identify the city s_j succeeding to j by following the tour in the direction $i \rightarrow s_i$.

In the same way, for the 3-opt neighborhood, we can quickly detect three cities i, j, and k that are close and can be candidates for a 3-opt move. But we cannot determine in a constant time if, starting from i, the tour visits first the city j before the city k. Indeed, if the 3-opt move (i, j, k) is feasible for a solution, the move (i, k, j) is not: it creates three sub-tours.

5.3 Neighborhood Extension

There are problems for which it is challenging to imagine reasonable neighborhoods which substantially modify a solution. Hence, the following problem arises: how to implement substantial changes to a solution on the basis of a simple and limited neighborhood.

Let us remark there is no contradiction in both limiting the size of a simple neighborhood with the techniques described above (to eliminate the moves that will never lead to good solutions) and extending this limited neighborhood with the techniques presented in this section.

5.3.1 *Filter and Fan*

To construct a neighborhood N_e extended from the definition of a small set M of moves applicable to a solution, we can consider k successive modifications $N_e(s) = \{s'|s' = s \oplus m_1 \oplus \cdots \oplus m_k, m_1, \dots m_k \in M(s)\}$. The size of N_e increases exponentially with k. To avoid such a growth, we can use the beam search strategy presented in Sect. 4.4.1, but adapted to a local search rather than a constructive method.

This technique, proposed by Glover [4], is called the *filter and fan* strategy. Each level only retains the p best neighbor solutions. Few of them may be worse than the starting solution. Then their neighbors are evaluated before repeating the process up to level k. Thus, at each step, up to k modifications are made according to the original neighborhood. It should be noted here that the best solution encountered when evaluating the extended neighborhood is not necessarily one of the ultimate level. This process is illustrated in Fig. 5.9.

A variant of filter and fan search is not to retain a static number p of neighbor solutions at each level, but all the improving neighbor solutions. The choice of the ultimately retained neighbor solution at level 1 is not the one that leads to the best solution up to level k, but the one which most opens the fan, that is to say, the solution of level $k - 1$ which has the most improving solutions at level k.

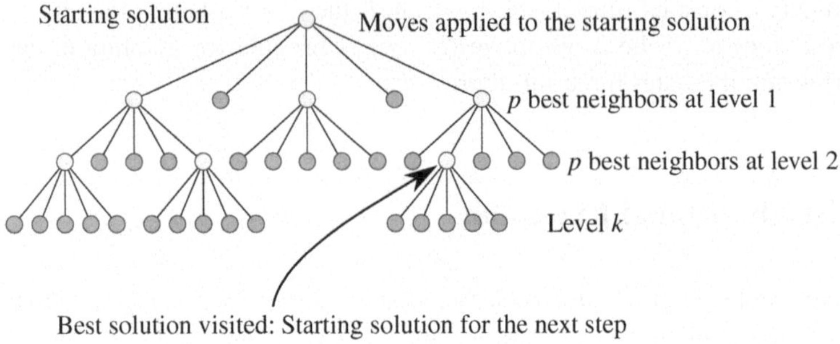

Fig. 5.9 Filter and fan with $p = 3$ and $k = 3$. Each level retains p neighbor solutions at most. The next current solution is the best of those listed

5.3.2 Ejection Chain

Another way to build a large neighborhood from basic modifications is to go through unfeasible solutions. The name of *ejection chain* has been proposed by Glover [3]. It combines and generalizes ideas from different sources. The most famous is certainly the Lin and Kernighan neighborhood for the TSP.

A starting solution is transformed into an object called a *reference structure*. The last is not a proper solution, but it can easily be transformed either into other reference structures or into feasible solutions. The starting solution is disrupted by the ejection of one of its components to obtain a reference structure which can also be transformed by the ejection of another component. This chain of ejections ends either when a better solution than the starting one has been identified or when all the elements to eject have been tested.

If an improving solution is discovered, the process is reiterated from it. Otherwise, the chain is initiated by trying to eject another item from the initial solution. The process stops when all possible chain initializations have been vainly tried. To prevent an endless process, it is forbidden either to add an item previously ejected to the reference structure or to propagate the chain by ejecting an element that was added to the reference structure.

5.3.2.1 Lin-Kernighan Neighborhood

One of the most effective neighborhoods for the TSP is due to Lin and Kernighan [7]. It is based on an ejection chain. The initial tour is transformed into a path by removing an edge $[a, b]$. This path is transformed into a reference structure. The last consists of a path linked to a cycle by including an edge $[b, d]$. The removal of the edge $[c, d]$, which is part of the cycle of the reference structure, transforms the reference structure into another path.

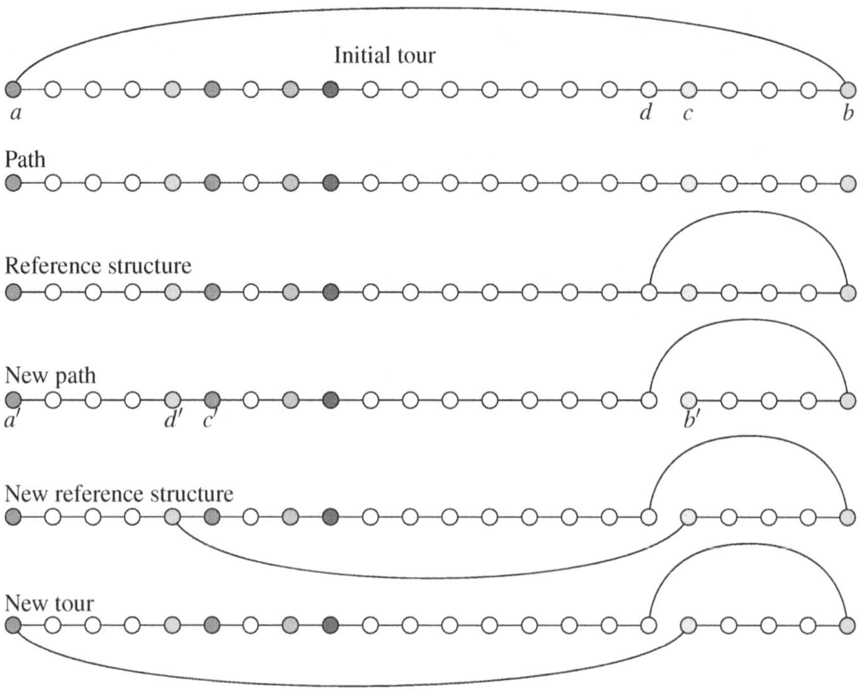

Fig. 5.10 Operating principle of an ejection chain for the TSP. Eject an edge $[a, b]$ to get a path. Insert an edge $[b, d]$ to get a reference structure. It can be transformed either into another reference structure, by ejection of the edge $[c, d]$ and addition of another edge, or into a tour by ejection of $[c, d]$ and addition of the edge $[a, c]$

The last is either transformed into a tour by adding the edge $[a, c]$ or in another reference structure by adding another edge incident to c. This node then plays the role of b of the previous step. Figure 5.10 illustrates the operating principle of this process.

The Lin and Kernighan local search is presented in Algorithm 5.4.

The ejection chain mechanism may seem artificial. However, it is possible to obtain relatively complex modifications and improve not awful solutions, as illustrated in Fig. 5.11.

A basic Lin and Kernighan implementation is given in Code 12.3. Much more elaborated, highly efficient implementations are due to [1] and [6].

Algorithm 5.4: Ejection chain (Lin and Kernighan) for the TSP

```
Input: TSP Solution s
Result: Improved solution s
repeat
    Eject edge [a,b] to initiate the chain
    repeat
        Find the edge [b,d] to add, minimizing the reference structure weight, with [c,d]
          not has been removed in the current ejection chain
        if Edge [b,d] found and reference structure weight smaller than tour s then
            if Adding [a,c] and removing [c,d] improve the tour then
                Success: Replace solution s by the new discovered tour
            else
                Add edge [b,d]
                Remove edge [c,d]
                b ← c
        else
            Ejection failure: come back to s and try another ejection
    until s improved or no edge [b,d] exists
until all edges of s have vainly initiated a chain
```

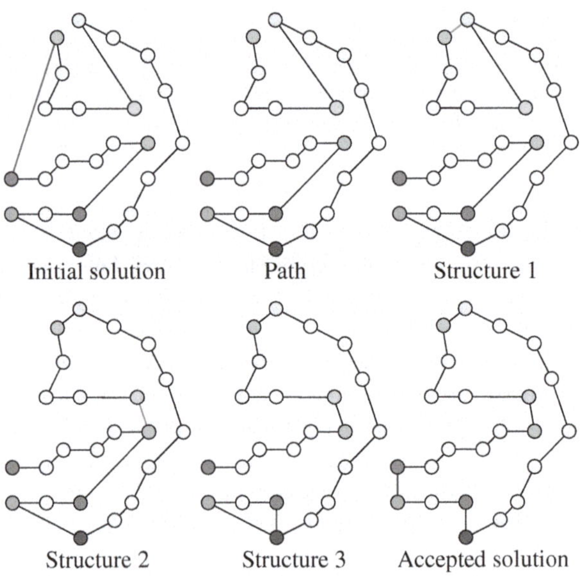

Fig. 5.11 Application of an ejection chain on a small TSP. After four ejections, it is possible to improve the starting solution

5.4 Using Several Neighborhoods or Models

A local optimum is relative to a given neighborhood structure. Hence, it is possible to use multiple neighborhoods simultaneously. For instance, a 2-optimal TSP tour is not necessarily 3-optimal.

Once a 2-optimum solution has been found, it is potentially possible to improve it with a method using a 3-opt neighborhood (see, for instance, Fig. 9.4). Similarly, a 3-optimal solution is not necessarily 2-optimal. We can therefore repeat the improvement processes as long as the solution found is not a local optimum with respect to all the neighborhood structures considered. The reader can verify this fact by running Code 12.7.

Finally, let us mention that one can switch from one modeling of the problem to another. To the extent that the neighborhood structure is not the same for the various modeling, it is equally possible to iterate improvement methods using different models. This technique may be inapplicable as is, since a feasible solution for one model can be unfeasible for another. In this case, repair methods should be provided when changing the modeling. This implies the process is no longer a strict improvement method. A corollary is that the search could enter an infinite cycle. Indeed, repairing a solution obtained with a first model and then improving it with a second model can cancel out the improvements obtained with the first model.

5.5 Multi-Objective Local Search

Various relatively simple local search techniques have been proposed for multi-objective optimization. We will examine two approaches not so difficult to implement and producing good solutions.

5.5.1 Scalarizing

A technique mentioned in Sect. 3.2.1 for multi-objective optimization is scalarizing. It aggregates the objectives by associating a weight w_i with the ith one. By providing a vector of weights and transmitting a scalar function to a local search, we can thus get an approximation of a supported solution. By varying the weight vector, we can discover more of these approximations. However, this technique produces at best one solution approximating the Pareto set for each local search run.

Without needing much more programming efforts, it is possible to get a better approximation of the Pareto set by transmitting all the objectives to the local search. Hence, we can check each neighbor solution whether it improves the Pareto set approximation. An elementary implementation of this principle is presented

Algorithm 5.5: Framework of an improvement method for multi-objective optimization. The user must provide a parameter I_{max} giving the number of scalarizations. Here, the weights are just randomly generated

Input: Solution s, neighborhood $N(\cdot)$, objective functions $\overrightarrow{f}(\cdot)$ to minimize; parameter I_{max}

Result: Approximation P of the Pareto set

$P = s$
for I_{max} *iterations* **do**
  Randomly draw a weight vector $\overrightarrow{w}$
  repeat
    $end \leftarrow$ true
    $best\_neighbor\_value \leftarrow \infty$
    forall $s' \in N(s)$ **do**
      if *s' is not dominated by solutions of P* **then**
        Insert s' in P and remove the solutions of P dominated by s'
      if $\overrightarrow{w} \cdot \overrightarrow{f}(s') < best\_neighbor\_value$ **then**
        $best\_neighbor\_value \leftarrow \overrightarrow{w} \cdot \overrightarrow{f}(s')$
        $best\_neighbor \leftarrow s'$
    if $best\_neighbor\_value < \overrightarrow{w} \cdot \overrightarrow{f}(s)$ **then**
      $s \leftarrow best\_neighbor$
      $end \leftarrow$ false
  until *end*

in Algorithm 5.5. Figure 5.12 illustrates the behavior of Algorithm 5.5 for three different scalarizations.

5.5.2 *Pareto Local Search*

An alternative approach to scalarization is the *Pareto local search* [10]. The idea is to start with any solution to the problem. The last is the first estimate—of poor quality—of the Pareto set. While the estimated Pareto set is not stabilized, generate all the neighbor solutions of the estimated Pareto set and update it with them.

Recursive procedures allow expressing the method very concisely, as shown by Algorithm 5.6. Code 5.5 implements a Pareto local search for the TSP.

However, the method can be sped up by not starting with an arbitrary solution, but by calling it several times with good solutions obtained by scalarizing the objectives. Indeed, there are habitually very effective methods for solving mono-objective problems. This allows us to immediately get supported solutions.

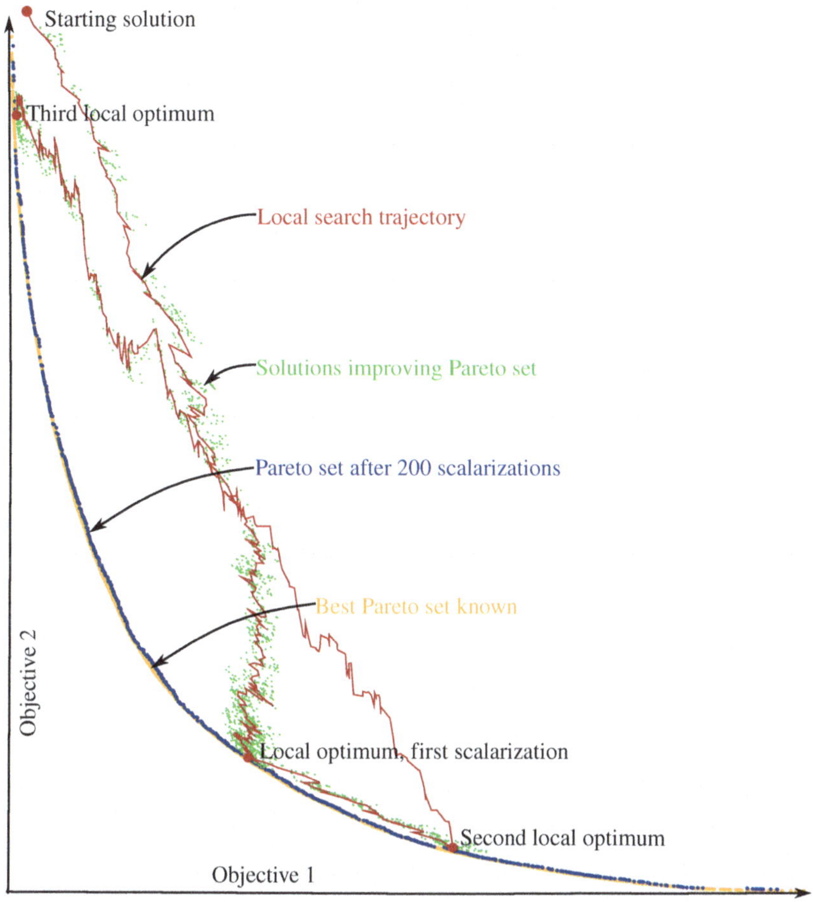

Fig. 5.12 Trajectory and evolution of an iterated local search for the bi-objective TSP instance (EuclidAB100) using the scalarization technique. The improvement method is based on an ejection chain. The weights associated with the objectives are randomly drawn each time the search reaches a local optimum. The starting solution is obtained by a greedy constructive method working solely on the first objective

For instance, polynomial algorithms exist for the linear assignment problem, while the multi-objective version is NP-hard. Starting Pareto local search with a reliable estimate of the Pareto set limits the number of updates, and the procedure stops faster. This also avoids too deep recursive calls, likely overflowing the recursion stack.

Algorithm 5.6: Framework of Pareto local search for multi-objective optimization. The interest of the method is to contain no parameter

Neighborhood_evaluation
  Input: Solution s; neighborhood $N(\cdot)$ objective functions $\overrightarrow{f}(\cdot)$
  Result: Pareto set P completed with neighbors of s
forall $s' \in N(s)$ **do**
  Update_Pareto($s', \overrightarrow{f}(s')$)
Update_Pareto
  Input: Solution s, objective values $\overrightarrow{v}$
  Result: Updated Pareto set P
if $(s, \overrightarrow{v})$ *either dominates a solution of P or $P = \varnothing$* **then**
  From P, remove all the solutions dominated by $(s, \overrightarrow{v})$
  $P \leftarrow P \cup (s, \overrightarrow{v})$
  Neighborhood_evaluation(s)

Code 5.5 tsp_3opt_pareto.py Implementation of Pareto local search for the TSP, based on 3-opt moves. This function calls another one (Code 12.6) that updates the Pareto set for each neighbor of the provided solution. The data structure used to store the Pareto set is discussed in Sect. 5.5.3

```
######### Pareto local search for the TSP based on 3-opt neighborhood
def tsp_3opt_pareto(pareto,                                    # Pareto front
                    costs,                        # Tour length (for each dimension)
                    s,                            # Solution (successor of each city)
                    d):                    # Distance matrix (one for each dimension)
    from kd_tree_update_pareto import update_3opt_pareto          # Listing 12.6
    from kd_tree_add_scan import K                                # Listing 12.4
    from random_generators import unif                            # Listing 12.1
    costs_neighbor = [-1 for _ in range(K)]            # Cost of neighbor solution
    start = unif(0, len(s)-1)                     # Starting city for move evaluation
    i, j, k = start, s[start], s[s[start]]              # Indices of a 3opt move
    while True:
        for dim in range(K):
            costs_neighbor[dim] = costs[dim] \
               + d[dim][i][s[j]] + d[dim][j][s[k]]  + d[dim][k][s[i]] \
               - d[dim][i][s[i]] - d[dim][j][s[j]]  - d[dim][k][s[k]]
        s[i], s[j], s[k] = s[j], s[k], s[i]        # Change solution to neighbor
        pareto = update_3opt_pareto(pareto, costs_neighbor, s, d)
        s[k], s[j], s[i] = s[j], s[i], s[k]                  # Back to solution
        k = s[k]                                                       # Next k
        if k == i:                                   # k at its last value, next j
            j = s[j]; k = s[j]
        if k == i:                                   # j at its last value, next i
            i = s[i]; j = s[i]; k = s[j]
        if s[s[i]] == start:                     # Neighborhood completely evaluated
            break
    return pareto
```

5.5.3 *Data Structures for Multi-Objective Optimization*

Both techniques presented above for multi-objective optimization can be relatively inefficient if no adequate data structure is used to store the Pareto set. Indeed, it often requires a constant time to evaluate the objectives of a neighbor solution. This means a few nano-seconds on current computers. Checking for the domination of a solution can consume much more time than computing the objectives. Using a simple list to store the p solutions of the Pareto set may slow down the search by a factor proportional to p.

It is not uncommon for the Pareto set to contain thousands of solutions. Hence, an appropriate data structure for testing the dominance of a solution should not require a computational time growing linearly with the Pareto set size.

5.5.3.1 Array

Assuming a limited number of different integer values for the K objectives, a $K - 1$ dimensional array is a simple and extremely efficient data structure to store the Pareto set.

The size of this array in a dimension is given by the distinct possible values the corresponding objective can take. A cell of this array stores the value of the best solution found for the Kth objective. For a bi-objective problem, we have a simple array. For instance, if we know that the first objective can vary from 2 to 13 and we have identified solutions with objectives $(2, 27)$, $(4, 24)$, $(6, 23)$, $(7, 21)$, and $(11, 17)$, the array contains $[27, 27, 24, 24, 23, 21, 21, 21, 21, 17, 17, 17]$. After the discovery of the solution $(5, 20)$, the array is updated as follows:
$[27, 27, 24, 20, 20, 20, 20, 20, 20, 17, 17, 17]$.

This data structure is limiting, because the objectives are not necessarily integers or do not involve a reasonable number of different values. However, if this data structure is usable in practice, it is incomparably fast, since it is possible to know the domination status of a solution in constant time.

5.5.3.2 KD-Tree

In the general case, a data structure whose query time is weakly growing with the number p of elements stored is the *KD-tree*. It is a binary search tree, where a node at the depth d discriminates the other elements of the tree on the dimension d mod K. Code 12.4 presents the basic procedures for including a new element in a KD-tree and inspecting all the elements stored in the tree.

The removal of a given node is a tricky procedure to program for a KD-tree. Figure 5.13 gives an example of updating a KD-tree after the discovery of a new efficient solution. Unlike a single-dimensional binary tree, the rightmost node from the left subtree or the leftmost one of the right subtree can generally not replace the removed node. Indeed, a KD-tree discriminates on a different dimension at each level. So, a walk through both sub-trees is required to find the replacing node. The last is itself recursively replaced, until it is a leaf, simply eliminated.

Code 12.5 implements a removal procedure of a given node within a KD-tree. Finally, to use a KD-tree to update a Pareto set, a function must find a point of the

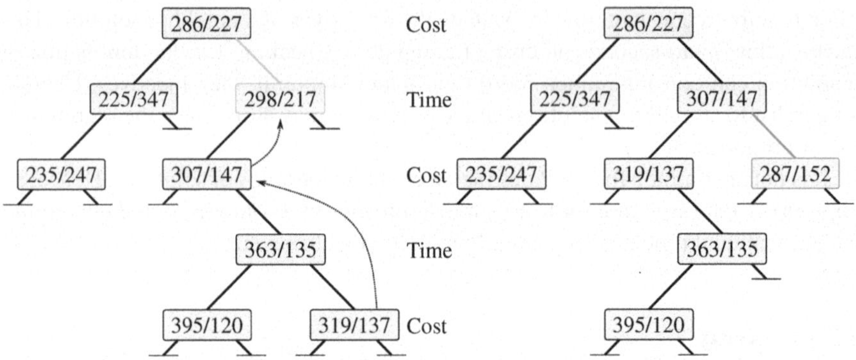

Fig. 5.13 KD-tree update for a bi-objective cost/time problem. On the left, the tree contains an approximation of the Pareto frontier. After the discovery of the efficient solution 287/152, the dominated solution 298/217 must be searched and eliminated. To do this, we look for the solution with the highest time in its left subtree (307/147). The last replaces the node of the dominated solution. The old node 307/147 must be recursively replaced, here, by the node (319/137) with the lowest cost in its right subtree (because the left subtree is empty). On the right is the situation of the KD-tree after insertion of the new efficient solution

tree that dominates an attempted solution, if any. For this, it is necessary to look if the KD-tree possesses a point between the ideal point, which is the one whose values on all dimensions are those of the optimum of the objectives, taken separately.

In practice, the ideal point is unknown, but a coarse approximation is appropriate, for instance, $(-\infty, \ldots, -\infty)$, if all the objectives must be minimized. If the KD-tree contains points in the hyper-rectangle delimited by $(-\infty, \ldots, -\infty)$ and the attempted solution, then these points dominate the trial solution. The last is thus ignored. Otherwise, the trial solution dominates others, which must be eliminated from the KD-tree. The trial solution is then added to the KD-tree. Code 12.6 allows updating a Pareto set when seeking to include a new point.

Problems

5.1 Local Minima

Figure 5.2 shows the local optima of a function of a discrete variable x relative to a neighborhood consisting in changing x by one unit. On this figure, locate the local optima relative to an asymmetric neighborhood consisting in either adding 4 or subtracting 3 from x. Does this neighborhood have the connectivity property?

5.2 Minimizing an Explicit Function

An integer function of integer variables $[-7, 6] \times [-6, 7] \to [-10, 650]$ is explicitly given in Table 5.1. We seek the minimum of this function by applying a local search with the first improvement policy, starting from the solutions (6, 7) of value 650 and

Table 5.1 Integer function $f(x, y)$ explicitly given

		x													
		−7	−6	−5	−4	−3	−2	−1	0	1	2	3	4	5	6
y	7	248	216	210	222	230	234	256	304	336	372	428	495	585	650
	6	193	175	157	166	174	181	215	249	295	329	382	454	539	597
	5	138	144	126	116	124	150	184	194	250	305	361	425	480	566
	4	123	89	85	97	105	109	129	179	209	246	302	368	458	525
	3	92	58	70	70	78	94	98	148	168	223	282	339	413	510
	2	68	34	46	46	54	70	74	124	144	199	258	315	388	486
	1	51	17	14	25	33	38	57	107	136	174	230	296	386	454
	0	18	25	5	−4	3	29	65	74	131	185	240	305	361	445
	−1	27	6	−10	0	8	13	46	83	126	160	213	284	371	429
	−2	33	0	−3	8	15	20	39	89	118	156	212	278	368	436
	−3	33	12	−4	6	14	19	52	89	132	166	219	290	377	435
	−4	30	37	17	7	15	41	77	86	143	197	252	317	373	457
	−5	69	35	32	43	51	56	75	125	154	192	248	314	404	472
	−6	92	58	70	70	78	94	98	148	168	223	282	339	412	510

from the solution $(6, -6)$ of value of 510. It is assumed that the moves consist in modifying by a unit the value of a variable. The moves are checked in the order: $(+1, 0)$, $(0, +1)$, $(-1, 0)$, and $(0, -1)$. Next, apply a local search with the best improvement policy, starting from the solution $(-7, 7)$ of value 248 and from the solution $(-7, -6)$ of value of 92.

5.3 2-Opt and 3-Opt Neighborhood Properties

Show the following properties of the 2-opt and 3-opt neighborhoods:

- The inversion of two cities or a 3-opt move can be obtained by a succession of 2-opt moves.
- 2-opt and 3-opt neighborhoods have connectivity property.

Provide an upper bound to the diameter of these neighborhoods.

5.4 3-Opt for Symmetric TSP

Section 5.1.3.1 introducing the 3-opt neighborhood shows a possibility to replace three arcs with three other arcs. This possibility respects the direction of travel of the three sub-paths located between the modified arcs. In the case of a symmetric problem where one accepts to change the direction of travel of some sub-paths, how many possibilities are there to replace three edges with three other edges while keeping an admissible tour?

5.5 4- and 5-Opt
For an asymmetric TSP, how many ways are there to replace four arcs with four other arcs while maintaining the direction of travel of the sub-paths? Same question for replacing five arcs.

5.6 Comparing 2-Opt Best and First
How many moves should be tested to show that a TSP tour with n cities is 2-optimal? Empirically evaluate the number of repetitions of the external loop. Provide this number as a function of the problem size for both procedures `tsp_2opt_first` and `tsp_2opt_best`. Analyze the difference if the procedures start either with the nearest neighbor solution or with a random one. Consider examples of Euclidean problems, randomly, uniformly generated in a square. Explain the results.

5.7 3-Opt Candidate List
To limit the size of a TSP neighborhood, only the 40 shortest arcs incident to each city are considered. With such a limitation, what is the complexity of verifying if a solution is 3-optimal? Is a special data structure required to achieve this minimum computational complexity? Is the neighborhood generated by such a limitation connected?

5.8 VRP Neighborhoods
Suggest four different neighborhoods for the vehicle routing problem. Give the size of these neighborhoods as a function of the number n of customers and the number m of tours. Specify whether these neighborhoods have connectivity property, depending on the problem modeling.

5.9 Steiner Tree Neighborhood
Two solution modelings have been proposed in Sect. 2.1.2 for the Steiner tree problem. Suggest neighborhoods adapted to each of these modelings.

5.10 Ejection Chain for the VRP
Propose a technique based on ejection chains for the vehicle routing problem. Specify how to initialize the chain, how to propagate it, and how to stop it. Estimate the computational complexity of evaluating an ejection chain for a solution with n customers and m tours.

References

1. Applegate, D.L., Bixby, R.E., Chvátal, V., Cook, W.J.: Concorde: A code for solving Traveling Salesman Problems. https://github.com/matthelb/concorde (1999). Accessed 16 June 2022
2. Croes, G.A.: A method for solving traveling salesman problems. Oper. Res. **6**, 791–812 (1958). https://doi.org/10.1287/opre.6.6.791

3. Glover, F.: Ejection Chains, reference structures and alternating path methods for traveling salesman problems. Discrete Appl. Math. **65**, 223–253 (1996). https://doi.org/10.1016/0166-218X(94)00037-E
4. Glover, F.: A template for scatter search and path relinking. In: Hao, J.K., Lutton, E., Ronald, E., Schoenauer, M., Snyers, D. (eds.) Artificial Evolution, Lecture Notes in Computer Science, vol. 1363, pp. 13–54 (1998). https://doi.org/10.1007/BFb0026589
5. Helsgaun, K.: An effective implementation of the Lin-Kernighan traveling salesman heuristic. Eur. J. Oper. Res. **126**(1), 106–130 (2000). https://doi.org/10.1016/S0377-2217(99)00284-2
6. Helsgaun, K.: Using POPMUSIC for Candidate set generation in the Lin-Kernighan-Helsgaun TSP solver. Department of Computer Science, Roskilde University, Denmark (2018)
7. Lin, S., Kernighan, B.W.: An effective Heuristic algorithm for the traveling-salesman problem. Oper. Res. **21**(2), 498–516 (1973). https://www.jstor.org/stable/169020
8. Or, I.: Traveling salesman-type combinatorial problems and their relation to the logistics of regional blood banking. Ph.D. Thesis, Northwestern University (1976)
9. Osterman, C., Rego, C.: A k-level data structure for large-scale traveling salesman problems. Ann. Oper. Res. **244**(2), 1–19 (2016). https://doi.org/10.1007/s10479-016-2159-7
10. Paquete, L., Chiarandini, M., Stützle, T.: Pareto local optimum sets in the biobjective traveling salesman problem: An experimental study. In: Gandibleux, X., Sevaux, M., Sörensen, K., T'kindt, V. (eds.) Multiobjective Optimisation. Lecture Notes in Economics and Mathematical Systems, vol. 535, pp. 177–200. Springer, Berlin (2004). https://doi.org/10.1007/978-3-642-17144-4_7
11. Reinelt, G.: TSPLIB — A T.S.P. Library . http://comopt.ifi.uni-heidelberg.de/software/TSPLIB95 (1990). Accessed 16 June 2022
12. Toth, P., Vigo, D.: The granular tabu search and its application to the vehicle-routing problem. INFORMS J. Comput. **15**(4), 333–346 (2003). https://doi.org/10.1287/ijoc.15.4.333.24890
13. Xavier, A.E., Xavier, V.L.: Flying elephants: A general method for solving non-differentiable problems. J. Heuristics. **22**(4), 649–664 (2016). https://doi.org/10.1007/s10732-014-9268-8

Chapter 6
Decomposition Methods

In the process of developing a new algorithm, this chapter should logically have been placed just after the one devoted to problem modeling. But, decomposition methods are only used when the data size to process is large. Thus, the phase is optional. The reader can glance it over before moving on to the following parts, devoted to the stochastic and learning methods. This is the reason justifying its place at the end of the first part of this book, devoted to the essential ingredients of metaheuristics.

6.1 Consideration on the Problem Size

The algorithmic complexity, very briefly exposed in Sect. 1.2.1, aims to evaluate the computational resources necessary for running an algorithm according to the data size it has to treat. We cannot classify the problems—large or small—only by their absolute size: sorting an array of 1000 elements is considerably easier than finding the optimal tour of a TSP instance with 100 cities. The time available to obtain a solution is clearly important: the perception of what a large instance is might not be the same if we have to perform a real-time processing in a few microseconds or a long-term planning for which a 1-day computation is perfectly convenient. Very roughly, we can put NP-hard problem instances in the following categories:

Toy Instances Approximative size: $n \approx 10$. To ensure an algorithm works correctly, it is performed by hand. Another possibility is to compare its results to those of a method, easy to implement, but much less efficient. For instance, this can be an exhaustive enumeration of all solutions. Yet, we can empirically consider a computer is able to perform 10^9 elementary operations per second. If one has a time budget of this order of magnitude, one can consider an exhaustive enumeration for a permutation problem instance up to $n \approx 10$; for a binary variable problem, we have $n \approx 20$. Naturally, for polynomial algorithms, the instance size processed in one second varies from $n \approx 50$ for complexity in

É. D. Taillard, *Design of Heuristic Algorithms for Hard Optimization*, Graduate Texts in Operations Research, https://doi.org/10.1007/978-3-031-13714-3_6

$O(n^3)$ to $n \approx 10^8$ for linear complexity, passing through $n \approx 10^4$ for quadratic complexity, and $n \approx 10^6$ for an algorithm in $O(n \log n)$.

Small Instances Typical size: $10 \lessapprox n \lessapprox 10^2$. When the size no longer allows an exhaustive enumeration of all solutions, we go into the category of small instances. We could characterize them by those for which we know robust algorithms that allow getting an optimal solution in a reasonable time. It should be mentioned that the literature frequently reports exact algorithms for solving examples of "difficult" problems of much larger size than those mentioned above. However, one should be careful with such statements: indeed, optimal solutions of traveling salesman or knapsack instances with tens of thousands of elements have been found, but much smaller instances are out of the scope of these programs. Small instances are useful for designing and calibrating heuristic methods. Knowing the optimal solutions allows determining the quality of heuristics and tuning the value of their parameters while maintaining reasonable computational times.

Standard Instances Typical size: $10^2 \lessapprox n \lessapprox 10^4$. This is the typical application area of metaheuristics. These are frequently encountered in real-world applications. They are too large to be solved efficiently by exact methods or for a human to guess a good quality solution. The maximum instance size a metaheuristic can handle is related to its algorithmic complexity, whether in terms of computational time or memory. With more than 10^4 elements, it becomes challenging to use a constructive method or a neighborhood size in $O(n^2)$. This is specially the case if one has to memorize an $n \times n$ matrix for efficiency reasons. The algorithmic complexity of a metaheuristic-based program is frequently larger than $O(n^3)$. Thus, many authors speak of a "large" instance for a size of 100.

Large Instances Typical size: $10^3 \lessapprox n \lessapprox 10^8$. Some real instances often have a higher number of items than standard instances, or they must be solved with less computational effort than a direct method would take. We can think, for instance, to vehicle routing for mail delivery or item labeling on a geographic map. For such problems, a size of 10^5 is not exceptional. In this case, decomposition methods must be used. This chapter presents some general techniques for approaching large instances. Let us mention that these techniques sometimes can advantageously be applied to smaller instances, even with just a few dozen elements.

Huge Instances Size: $n > 10^8$ items. When the size of the problem exceeds 10^8 to 10^{10} items, it is no longer possible to completely store the data in RAM. In this case, it is necessary to work on parts of the instance, usually using parallel algorithms to maintain adequate processing times. The treatment of this type of instances essentially raises mainly technical issues and is beyond the scope of this book.

6.2 Recursive Algorithms

When a large instance has to be solved with limited computational effort, it is cut into small parts, independently solved. Finally, they are put together to reconstruct a solution to the complete problem. An efficiency gain is only possible with such a technique by the conjunction of several conditions: directly solving the problem requires a computational effort more than linear; otherwise, a decomposition only makes sense for a parallel computation. The parts must be independent of each other. Combining the parts together should be less complex than directly solving the problem. The difficulty lies in how to define the parts: they must represent a logical portion of the problem so that their assembly, once solved, is simple.

The merge sort is a typical decomposition algorithm. A list to sort is split into two roughly equal parts. These are sorted by two recursive calls, if they contain more than one element. Finally, two locally sorted sub-lists are scanned to reconstruct a complete sorted list.

6.2.1 Master Theorem for Divide-and-Conquer

In many cases, the complexity of a recursive algorithm can be assessed by the divide-and-conquer master theorem. Suppose the time to address a problem of size n is given by $T(n)$. The algorithm proceeds by splitting the data into b parts of approximately identical size, n/b. Among them, a are recursively solved. Next, these parts are combined to reconstruct a solution to the initial problem, which requires a time given by $f(n)$. To assess the complexity of such an algorithm, we must solve the functional equation $T(n) = a \cdot T(n/b) + f(n)$ whose solution depends on the reconstruction effort.

Introducing ϵ, a positive constant forcing the function $f(n)$ to be either smaller or larger than $n^{\log_b(a)}$, the master theorem allows deducing the complexity class of $T(n)$ in some case:

- If $f(n) = O(n^{\log_b(a)-\epsilon})$, then $T(n) = \Theta(n^{\log_b(a)})$.
- If $f(n) = \Theta(n^{\log_b(a)})$, then $T(n) = \Theta(n^{\log_b a} \cdot \log n)$.
- If $f(n) = \Omega(n^{\log_b(a)+\epsilon})$ and if $a \cdot f(n/b) < c \cdot f(n)$, with $c < 1$, constant, then $T(n) = \Theta(f(n))$.

Often, $a = b$: we have a recursive call for all parts. In this case, the theorem states that, if the reconstruction can be done in a sublinear time, then we can deal with the problem in linear time. If the reconstruction takes a linear time—which is typically the case for sorting algorithms—then the problem can be solved in $O(n \log n)$. The last case simply indicates all the difficulty of the algorithm is concentrated in the reconstruction operations. Finally, let us mention that the theorem does not cover all cases for the function $f(n)$.

There are also cases where $a \neq b$. An example is a query for a point of the Euclidean plane from a set of n points stored in a balanced 2D-tree (see the data structure discussed in Sect. 5.5.3.2). With such a data structure, one can halve the number of points remaining to be examined by processing a maximum of $a = 2$ parts among $b = 4$. Indeed, unlike a binary tree in one dimension, we cannot ensure to divide this number by two at every single level of the tree but only every two levels. Since this is a query problem, there are no reconstruction and $f(n) = O(1)$. As $\log_4(2) = 1/2$, we can choose $\epsilon = 1/2$, and we are in the first case. We can deduce that the complexity of a query in a 2D-tree is in $\Theta(n^{1/2})$. However, if the points are well spread, the empirical behaviour is better, closer to $\log n$.

Heuristic algorithms proceeding by recursion commonly stop prematurely, before the part size is so small that its resolution becomes trivial. Even if the parts are exactly solved, the reconstitution phase does not generally guarantee optimality. Hence, both cutting and reconstitution procedures are heuristics. This means that the "border areas" between two parts are, more or less obviously, not optimum. To limit this effect of sub-optimality, it is necessary to assemble as few parts as possible, while being able to process them. Indeed, if they are excessively large, their exact resolution requires too much time, or the heuristics may produce low-quality parts.

6.3 Low Complexity Constructive Methods

Solving large instances implies limiting the complexity of the constructive method for generating an initial solution. This means that even the most basic greedy method is not appropriate. If the function $c(s, e)$ that provides the cost of the addition of an element e actually depends on the partial solution s, then its complexity is in $\Omega(n^2)$. Indeed, before including one of the n elements, it is necessary to evaluate $c(s, e)$ for all the remaining elements. A random construction in linear time is not suitable, due to the bad quality of the solution produced.

It is therefore necessary to "cheat," making the hypothesis that not *all* the elements of the problems have a direct relationship with *all* the others. Put differently, an element is in relation with a relatively limited number of other elements, and this relationship possesses a certain symmetry. It is reasonable to make the hypothesis that it is possible to quantify the proximity between two elements. In such a case, we can avoid complexity in $O(n^2)$ by sampling and recursion. We can limit the phenomenon of sub-optimality due to the assembly of parts by stopping the recursion at the first or the second level.

6.3.1 Proximity Graph Construction

There are relatively good automatic classification heuristics to partition a problem of size n into k groups. The fast variant of Algorithm 2.7 (k-medoids) mentioned in Sect. 2.7.2 achieves such a heuristic partition with complexity of $\overline{O}(k \cdot n + (\frac{n}{k})^2)$.[1]

This complexity can be minimized by choosing $k = \sqrt{n}$. Thus, it is possible to partition a problem of size n in $\sqrt{n}$ parts, each comprising approximately $\sqrt{n}$ elements. Performing the clustering on a random sample of the elements (e.g., $\Theta(\sqrt{n})$) can significantly speed up the procedure. This method is illustrated in Fig. 6.1.

It is possible to get a decomposition with smaller clusters by applying a second recursion level: the instance is first cut into a large parts of relatively similar size as presented above. A proximity relationship is defined between large part, so that each includes $O(1)$ neighbors. A rudimentary proximity definition is as follows: if an element has c_i as its nearest center and c_j as its second nearest, then c_i and c_j are considered as neighbors. Each large part is then partitioned into b small clusters.

Similarly, a proximity relationship is defined between small clusters. A small cluster is related to all those belonging to the large part of which it belongs. By choosing $a = \Theta(\sqrt{n})$ and $b = \Theta(\sqrt{n})$, we get a decomposition into a number of small clusters proportional to n, whose size is approximately identical. The overall algorithmic complexity is $\overline{O}(n^{3/2})$.

For some problems, it can make sense. Indeed, for the vehicle routing problem, the maximum number of customers that can be placed on a tour depends on the application (home service, parcel distribution, rubbish collection) and not on the total number of customers of the instance.

This decomposition technique is illustrated in Fig. 6.2. Bold lines show proximity relations between large parts. The small clusters obtained by decomposition of large parts contain about 15 elements. The elements of large parts are represented by points of the same color. By exploiting such a decomposition and proximity relationships, it becomes possible to efficiently generate a solution to a large problem instance. A computational time of about one second was enough to obtain the structures of Fig. 6.2, with more than 16,000 entities.

[1] The notation $O(\cdot)$ cannot be used here because it is assumed that the k groups contain approximately the same number of elements. In the worst case, a few groups could contain $\Theta(n)$ elements, and $\Theta(n)$ groups could contain $O(1)$ elements, which would imply a theoretical complexity in $O(n^2)$. To limit the complexity, the algorithm must be stopped prematurely, if needed, by repeating a constant number of times the external loop.

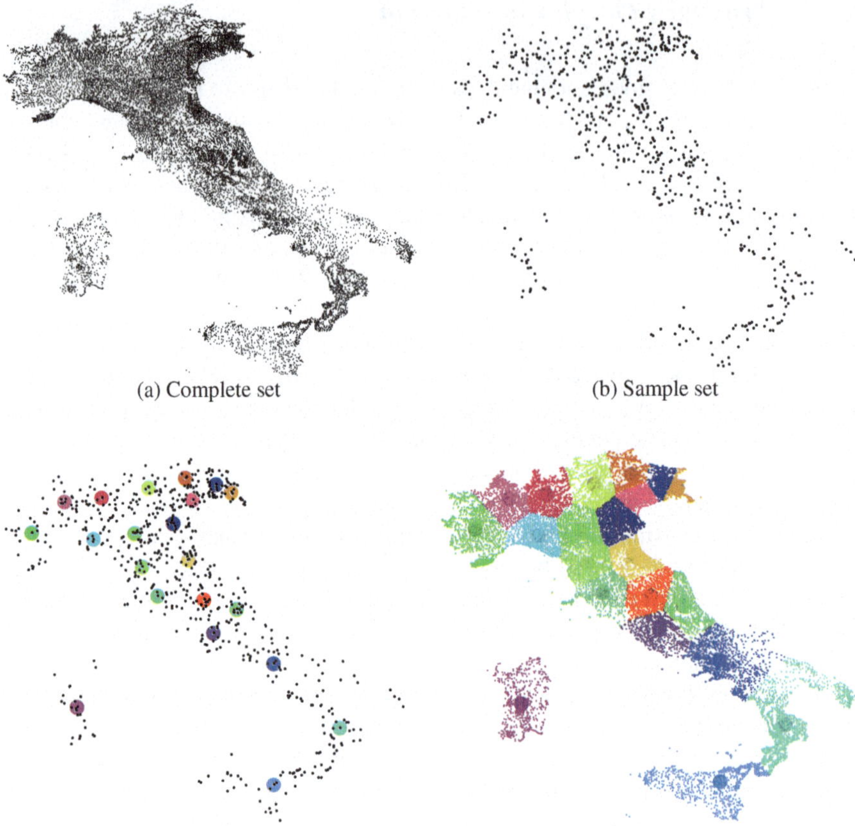

(a) Complete set (b) Sample set

(c) Centers (d) Clusters

Fig. 6.1 Illustration of the method for partitioning a problem instance: from a complete set of n elements of the instance (**a**), a random sample is selected (**b**). Algorithm 2.7 is run on the sample and $k = \Theta(\sqrt{n})$ medoids are identified (**c**). All the n elements are allocated to the closest medoid (**d**)

6.3.2 Linearithmic Heuristic for the TSP

It is possible to extend this decomposition principle to a number of levels depending on the instance size and thus get an $O(n \log n)$ algorithm. This section illustrates the principle on the Traveling Salesman Problem. Rather than reasoning on the construction of a tour, we build paths passing through all the cities of a given subset.

It is actually straightforward to adapt Code 12.3 so that it is able to treat a path rather than a tour. An algorithm to optimize a path can equally be used to provide a tour. Indeed, a TSP tour can be seen as a path starting by city $c_i \in C$ and ending by c_i. If we have a problem with n cities, the path $P = (b = c_i, c_1, \ldots, c_{i-1}, c_{i+1}, \ldots, c_n, c_i = e)$ defines a feasible (random) tour. The path

Fig. 6.2 Two-level decomposition of a problem instance. The elements are clustered into $\Theta(\sqrt{n})$ large parts of approximately identical size. Bold lines show the proximity relationship between large parts. The latter are themselves decomposed into $\Theta(\sqrt{n})$ small clusters. The complexity of the process is in $\overline{O}(n^{3/2})$. It can be applied to non-geometrical problems

P is either directly optimized if it does not contain too many cities, or decomposed into r sub-paths, where r is a parameter that does not depend on the problem size. To fix the ideas, the value of r is typically between 10 and 20. If $n \leq r^2$, a very good path passing through all the cities of P, starting in c_i and ending in c_i can be found, for example, with an ejection chain or even an exact method. This feasible tour is returned by the heuristic.

Else, if $n > r^2$, the path P is reordered by considering r sub-paths. This is performed by choosing a sample S of r cities by including:

- $u \in C \setminus \{b, e\}$, the city closest to b
- $v \in C \setminus \{b, e, u\}$, the city closest to e
- $r - 2$ other cities of $C \setminus \{b, e, u, v\}$ randomly picked

A good path P_S through all the cities of sample S, starting at city b and ending at city e, can be found with a local search or an exact method. Let us rename the cities

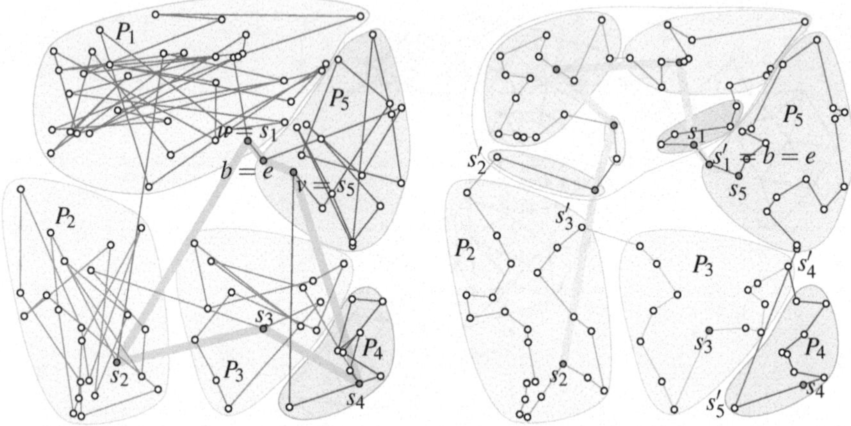

Fig. 6.3 Recursive TSP tour construction. Left: path on a random sample (bold and light line) and reordered path P_S completed with all cities before recursive calls of the procedure. Paths P_1 to P_5 are drawn with different colors and different backgrounds highlight them. Right: the path P_1 was recursively decomposed into $r = 5$ pieces. Final state with all sub-paths optimized

of S so that $P_S = b, s_1, s_2, \dots, s_{r-1}, s_r, e$. Path P_S can be completed to contain all the cities of C by inserting them, one after the others, just after the closest city of S. So, the completed path $P_S = (b, s_1, \dots, s_2, \dots, , s_t, \dots, e)$ improves the initial path P. The left side of Fig. 6.3 illustrates this construction using a sample of $r = 5$ cities. The shaded areas highlights the first r sub-paths found.

At this step, the order of the cities in the completed path P_S between two cities s_j and $s_j + 1$ is arbitrary (as it was for P at the beginning of the procedure). The sub-paths $P_1 = (b = s'_1, \dots, s_2)$, $P_2 = (s'_2, \dots, s_3), \dots, P_r = (s'_r, \dots, e = s_{r+1}) \subset P_S$ can be further improved with r recursive calls of the same procedure, where s'_j is the city just preceding the first one of the path P_j. The right side of Fig. 6.3 illustrates the solution returned by this recursive procedure.

It can be noted in this figure that only the sub-path P_1 has been decomposed. The others, not comprising more than r^2 cities, were directly optimized by a procedure similar to that given by Code 12.3. The solution finally obtained is not excellent, but it was obtained very quickly and is suitable as an initial solution for partial improvement techniques, like POPMUSIC, which will be detailed in Sect. 6.4.2.

6.4 Local Search for Large Instances

After reviewing some techniques for constructing solutions for large instances, let's now take a look at some techniques for improving them. LNS, POPMUSIC, and Corridor assume that an initial solution to the problem is available. These techniques are sometimes called fix-and-optimize [4] or, more recently, magnifying glass heuristics [3]. The key idea is to fix a relatively large portion of the problem variables and to solve a sub-problem with additional constraints on the remaining variables. When a heuristic includes an exact optimization method, we now speak of *matheuristic*.

6.4.1 Large Neighborhood Search

Large neighborhood search (LNS) has been proposed by Shaw [5]. The general idea is to gradually improve a solution by alternating destruction and repair phases. To illustrate this principle, let's consider the example of integer linear programming. The destruction phase involves selecting a subset of variables while incorporating some randomness into the process. In its simplest form, this consists in selecting a constant number of variables, in a completely random fashion. A more elaborate form is to randomly select a seed variable and a number of others, which are most related to the seed variable. The repair phase consists in trying to improve the solution by solving a sub-problem on the variables that have been selected. The value of the other variables being set to the one taken in the starting solution.

The name of this technique comes from the fact that a very large number of possibilities exist to reconstruct a solution. This number exponentially increases with the size of the sub-problem, meaning that they could not reasonably be extensively enumerated. Thus, the reconstruction phase consists in choosing a solution among a large number of possibilities. As the significant part of the variables preserves their value from one solution to the next, it is conceptually a local search but with a large neighborhood size. The framework of LNS is provided by Algorithm 6.1.

Algorithm 6.1: LNS framework. The destroy, repair, and acceptance functions must be specified by the programmer, as well as the stopping criterion

Input: Solution s, destroy method $d(\cdot)$, repair method $r(\cdot)$, acceptance criterion $a(\cdot,\cdot)$
Result: Improved solution s^*

$s^* \leftarrow s$
repeat
  $s' \leftarrow r(d(s))$
  if $a(s, s')$ **then**
    $s \leftarrow s'$
  if s' *better than* s^* **then**
    $s^* \leftarrow s'$
until *a stopping criterion is satisfied*

This frame leaves considerable freedom for the programmer to select various options:

Destroy method $d(\cdot)$ This method is supposed to destroy part of the current solution. The authors recommend that it is not deterministic, so that two successive calls destroy various portions. Another vision of this method is to fix a certain number of variables of the problem and release the others, which can be modified. This method additionally includes a parameter that allows modulating the amount of destruction. Indeed, if the number of independent variables is too small, the repair method has too many constraints to be able to differently reconstruct the solution, and the algorithm is not able to improve the current solution. Conversely, if the number of independent variables is too large, the repair method may encounter difficulties in improving the current solution. This is peculiarly true if an exact method is used, implying a prohibitive computational time.

Repair method $r(\cdot)$ This method is supposed to repair the part of a solution that was destroyed. Another vision of this method is to re-optimize the portion of the problem corresponding to the variables that were freed by the destroy method. One possible option for the repair method is to use an exact method, for instance, constraint programming. Another option is to use a heuristic method, either a simple one, like a greedy algorithm, or a more advanced one, such as taboo search, variable neighborhood search, etc.

Acceptance criteria $a(\cdot,\cdot)$ The simplest acceptance criterion is to use the fitness function value of both solutions provided as parameters:

$$a(s, s') = \begin{cases} True \text{ If } s' \text{ better than } s \\ False \text{ Otherwise} \end{cases}$$

Other criteria have been proposed, for instance, those inspired by simulated annealing (see Sect. 7.1).

Stopping criterion The framework does not provide any suggestion for the stopping criterion. Authors frequently use the limit of their patience, expressed in seconds. Also, it can be the patience of other authors who have proposed a concurrent method! This kind of stopping criteria is hardly convincing. This point is discussed further in Sect. 11.3.4.2. The quite close POPMUSIC framework, presented in Sect. 6.4.2, incorporates a natural stopping criterion.

To illustrate a practical implementation of this method, let us consider those of Shaw [5], originally adapted to the vehicle routing problem. The destroy method selects a seed client at random. The remaining customers are sorted using a function measuring the relationship with the seed customer. This function is inversely proportional to the distance between customers and depends on whether the customers are part of the same tour. The idea is to select a subset of customers who are close to the seed one but from different routes. These clients are randomly selected, with a bias to favor those most closely related to the seed client.

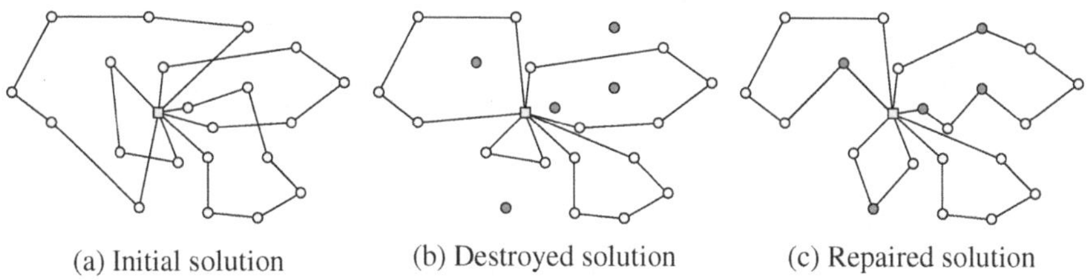

Fig. 6.4 Illustration of LNS on a VRP instance. The initial solution (**a**) is destroyed by removing a few customers (**b**). The destroyed solution is repaired by optimally inserting the removed customers (**c**)

The repair method is based on integer linear programming. The method implements a branch and bound technique with constraint propagation. This method can only modify the variables associated with the clients chosen by the destruction method. In addition, to prevent the explosion of computational times, common with exact methods, the enumeration tree is partially examined and heuristically pruned. A destroy-repair cycle is illustrated in Fig. 6.4.

There are algorithms based on the LNS framework that have been proposed well before it. Among these applications is the shifting bottleneck heuristic for the jobshop scheduling problem [1]. In this article, the destroy method selects the bottleneck machine and frees the variables associated with the operations processed by this machine. The repair method reorders these operations, considering that the sequences on other machines are not modified. Hence, each operation on the bottleneck machine has a release time corresponding to the earliest finishing time of the preceding operation on the same job. In addition, each operation has a due date, corresponding to the latest starting time of the following operation on the same job. In this heuristic, all choices are deterministic and all optimization are exact. So, the current solution is modified only if it is strictly improved and the method has a natural stopping criterion.

The POPMUSIC method presented in the following section was developed independently from LNS. It can be seen as a less flexible LNS method, in the sense that it better suggests to the programmer the choice of options, particularly the stopping criterion.

6.4.2 POPMUSIC

The primary idea of POPMUSIC is to locally optimize a part of an existing solution. These improvements are repeated until no part that can be optimized are detected. It is, therefore, a local search method. Originally, this method received the less attractive acronym of LOPT (for *local optimizations*) [8, 9].

For large problem instances, one can consider that a solution is composed of a number of parts, which are themselves composed of a number of items. Taking the example of clustering, each cluster can be a part. In addition, it is assumed that one can define a proximity measure between the parts and that the latter are somewhat independent of each other in the solution. In the case of clustering, there are closely related clusters, containing items that are not well separated, and independent clusters, that are clearly well separated. If these hypotheses are satisfied, we have the special conditions necessary to develop an algorithm based on the POPMUSIC framework. The name was proposed by S. Voß. It is the acronym of *Partial OPtimization Metaheuristic Under Special Intensification Condition*.

First, let us assume that a solution s can be represented by a set of q parts $s_1, \ldots, s_q$, and next that we have a method for measuring the proximity between two parts. The germinal idea of POPMUSIC is to select a seed part s_g and a number $r < q$ of the parts the nearest to s_g to build a sub-problem R. With an appropriate definition of the parts, improving the sub-problem, R can reveal an improvement for the complete solution. Figures 6.5 and 6.6 illustrate what a part and a sub-problem can be for various applications.

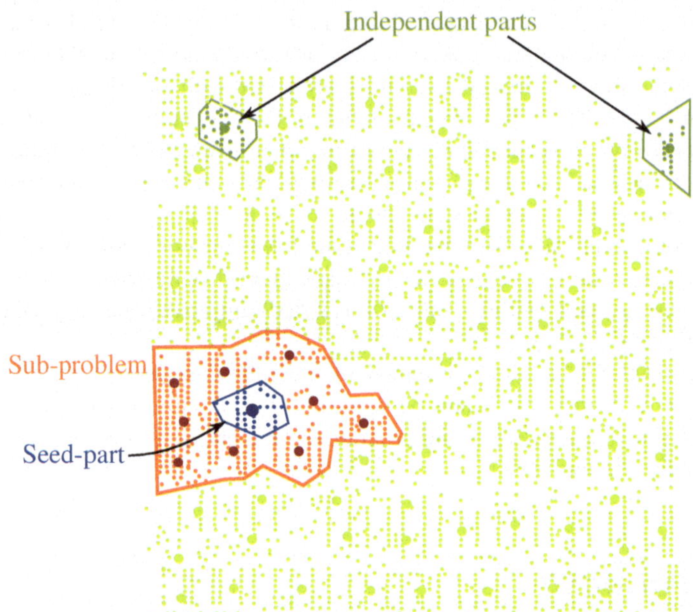

Fig. 6.5 To apply the POPMUSIC framework to a clustering problem, one can define a part as all the items assigned to the same center. The parts the nearest from the seed cluster constitute a sub-problem that is tentatively optimized independently. The optimization of well separated clusters cannot improve the solution. Hence, these parts are de facto independent

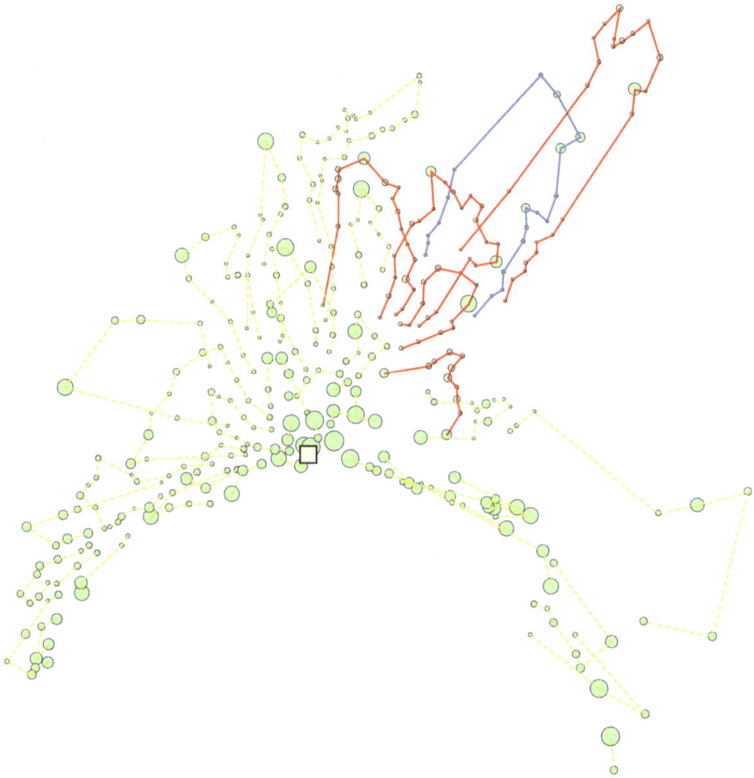

Fig. 6.6 For the VRP, the definition of a part in POPMUSIC can be a tour. Here, the proximity between tours is the distance of their center of gravity. A sub-problem consists of customers belonging to six tours

To prevent optimizing the same sub-problem several times, a set U stores the seed parts that can define a sub-problem potentially not optimal. If the tentative optimization of a sub-problem does not lead to an improvement, then the seed part used to define it is removed from U. Once U is empty, the process stops. If a sub-problem R has been successfully improved, a number of parts have been modified. New improvements become possible in their neighborhood. In this case, all parts of U that no longer exist in the improved solution are removed before incorporating all parts of R. Algorithm 6.2 formalizes the POPMUSIC method.

To transcribe this framework into a code for a given problem, there are several options:

Obtaining the initial solution POPMUSIC requires a solution before starting. The technique presented in Sect. 6.3 suggests how to get an appropriate initial solution with limited computational effort. However, POPMUSIC may also work for a limited instance size. In this case, an algorithm with a higher complexity can generate a starting solution.

Algorithm 6.2: POPMUSIC framework

Input: Initial solution s composed of q disjoint parts $s_1, \ldots, s_q$; sub-problem improvement method
Result: Improved solution s
$U = \{s_1, \ldots, s_q\}$
while $U \neq \varnothing$ **do**
  Select $s_g \in U$ // s_g: Seed part
  Build a sub-problem R composed of the r parts of s the closest to s_g
  Tentatively optimize R
  if *R is improved* **then**
    Update s
    From U, remove the part no longer belonging to s
    In U, insert the parts composing R
  else R not improved
    Remove s_g from U

Definition of a part The definition of a part is not unique for a given problem. In the VRP case, we can consider that all customers on the same tour form a part, as was done in [2, 7] (see Fig. 6.6). For the same problem, it is equally possible to define a part as a single client, as in [5].

Definition of the distance between parts For some problems, the definition of distance between two parts can be relatively easy and logical. For example, [9] uses the Euclidean distance between centroids for a clustering problem. For map labeling (Sect. 3.3.3), a graph is built whose vertices represent the objects to be labeled and the edges represent potentially incompatible label positions. The distance is measured by the minimum number of edges of a path to the seed label, as shown in Fig. 6.7.

By cons, this definition can be quite unclear for some problems. For the VRP with time window, two geometrically close clients can have incompatible opening time windows. Therefore, they should be considered as distant.

It is possible to use several different proximity definitions simultaneously. If we take the problem of school timetable design, one definition may aim to create sub-problems focusing on groups of students following the same curriculum, another on teachers, and a third on room allocation. Of course, if several definitions of proximity between parts are used simultaneously, the last line of Algorithm 6.2 has to be adapted: a seed part s_g will only be removed from U if none of the sub-problems that can be created with s_g can improve the solution.

Selection of the seed part To our knowledge, there are no comprehensive studies on the impact of the seed part selection process. In the literature, only very simple methods are used to manage the set U: either stack or random selection.

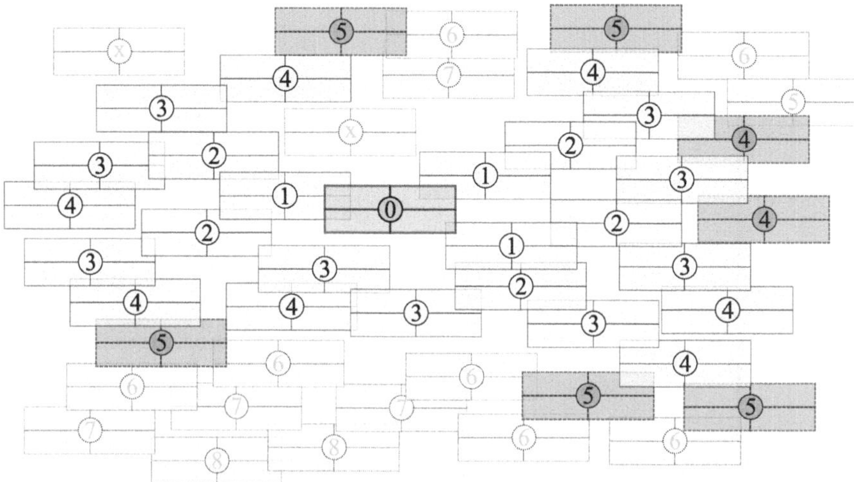

Fig. 6.7 For map labeling, a part can be an object to be labeled (circle with a number). Here, we consider four possible label positions (rectangles around each object). Two objects are at a distance of one if their labels may overlap. The number inside each disc represents the distance from the seed object, noted 0. A sub-problem has up to $r = 25$ objects which are the closest to the seed object. Here, the distance is at most 4. The objects whose labels could collide with these r objects are included in the sub-problem. Only the positions of the labels of the r objects can be changed when optimizing a sub-problem

Parameter r The size of the sub-problems depends on r, the only explicit parameter of POPMUSIC. It depends on the ability of the optimization method. A low value only allows minor improvements, but it requires a limited computational effort. A high value implies a high computational effort but a better potential to improve the solution.

Sub-problem optimization method The programmer is free to select any sub-problem optimization method. Since the sub-problem size can be adjusted, the implementation is facilitated: the method should be efficient for a limited span of instance size. In case the optimization method is an exact one, POPMUSIC framework is a matheuristic.

Looking at the stopping criterion—the set U is empty—the computational effort could potentially be prohibitive for large instances. Indeed, for each sub-problem improvement, several parts are introduced in U. In practice, the number of sub-problems to solve grows almost linearly with the instance size. Figure 6.8 illustrates this for a location-routing problem [2] and Fig. 6.10 for the TSP.

6.4.2.1 POPMUSIC for the TSP

An elementary implementation of the POPMUSIC technique for the traveling salesman problem is given by Code 6.1. In this adaptation, a part is a city. The

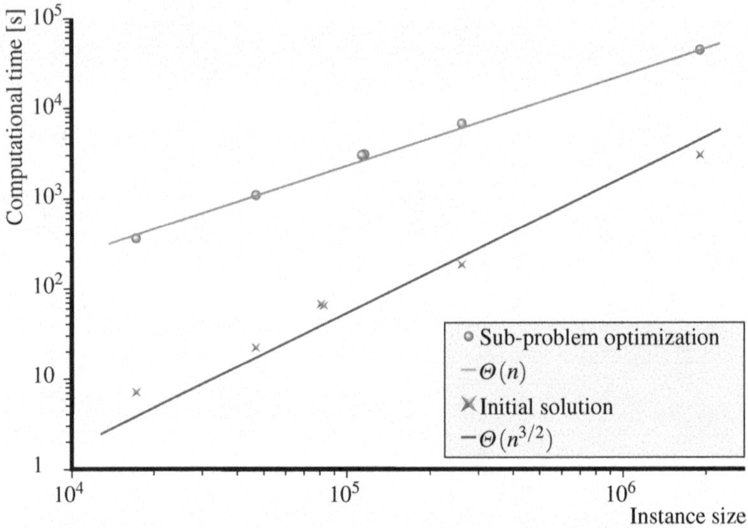

Fig. 6.8 Computational time observed for creating an initial solution to a location-routing problem with the technique presented in Sect. 6.3 and overall optimization time for sub-problems with the POPMUSIC frame. We notice that the growth of the computation time seems lower than the analysis of $\Theta(n^{3/2})$ done in Sect. 6.3 and that the time for the optimization of the sub-problems is almost linear

distance between parts is measured by the number of intermediate cities that there are along the current tour. This contrasts with a measure using the distance matrix. A sub-problem is, therefore, a path of $2r$ cities whose extremities are fixed. We seek to move a sub-path of at most r cities in the sub-problems, using a 3-opt neighborhood. The set U is not represented explicitly because it is identified to the tour. Indeed, successive sub-problems are just defined by a single city shift. To determine whether to continue to optimize, the initial city of the last sub-path that was successfully optimized is stored. If all starting cities are tried without improvement, the process stops.

Code 6.1 tsp_3opt_limited.py Basic POPMUSIC implementation for the TSP

```
######### POPMUSIC for the TSP based on 3-opt neighborhood
def tsp_3opt_limited(d,                                    # Distance matrix
                     r,                                    # Subproblem size
                     succ,                          # Tour provided and returned
                     length):                                   # Tour length
    n = len(succ)
    if r > n - 2:                    # Subproblem size must not exceed n - 2
        r = n - 2
    i = last_i = 0                                # starting city is index 0
    while True:
        j = succ[i]
        t = 0
        # do not exceed subproblem and the limits of the neighborhood
        while t < r and succ[succ[j]] != last_i:
            k = succ[j]
            u = 0
            while u < r and succ[k] != last_i:
                delta = d[i][succ[j]] + d[j][succ[k]] + d[k][succ[i]] \
                       -d[i][succ[i]] - d[j][succ[j]] - d[k][succ[k]]
                if delta < 0:                      # Is there an improvement?
                    length += delta                             # Perform move
                    succ[i], succ[j], succ[k] = succ[j], succ[k], succ[i]
                    j, k = k, j                  # Replace j between i and k
                    last_i = i
                u += 1
                k = succ[k]                                           # Next k
            t += 1
            j = succ[j]                                               # Next j
        i = succ[i]                                                   # Next i

        if i == last_i:          # A complete tour scanned without improvement
            break

    return succ, length
```

In order to successfully adapt the POPMUSIC technique to the TSP, it is necessary to pay attention to some issues:

- The initial solution must already possess an appropriate structure; for a Euclidean problem, it should not include two intersecting edges belonging to portions of routes that are separated by a long sequence of cities, because the optimization procedure will be unable to uncross them.
- Rather than developing an ad hoc local search like the one in Code 6.1 to optimize sub-paths, it is easier to use a general TSP solving method, for instance, Code 12.3.
- Ultimately, we must avoid optimizing a second time a sub-path that was already optimized.

To start with a solution having an appropriate structure, without using an algorithm of high complexity, we can go along the lines of the technique presented in Sect. 6.3.2. As the empirical complexity of POPMUSIC is linear, one can obtain a solution of satisfactory quality in $n \log n$ [10]. In practice, the time to build an

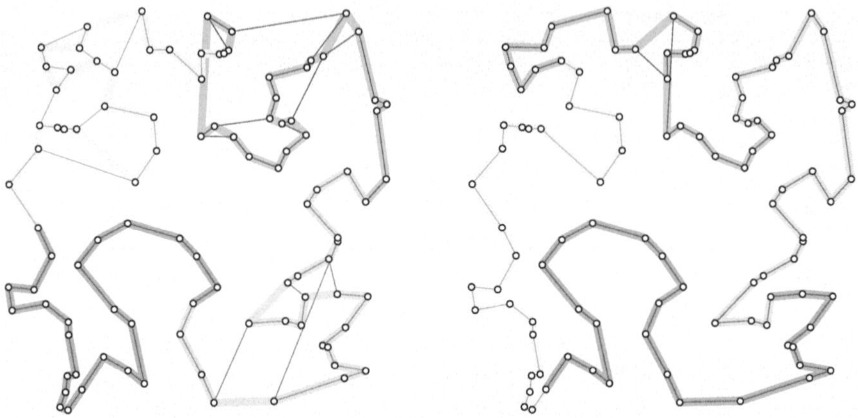

Fig. 6.9 On the right, independent optimizations of four sub-paths. The bold lines highlight the tour after optimization. The thin lines are those of the initial tour. On the left, the tour is shifted and the process is repeated

initial solution is negligible compared to its improvement with POPMUSIC, even for instances with billions of cities. We can speed up the process as follows, without significantly degrading the final solution: the route over n cities is cut into $\lceil n/r \rceil$ sub-paths of approximately r cities. These sub-paths are connected only by their extremities. Therefore, they can be independently optimized.

Once all these paths have been optimized, the tour is shifted by $r/2$ cities. Finally, $\lceil n/r \rceil$ sub-paths overlapping the previous ones are optimized. Thus, with $2 \cdot \lceil n/r \rceil$ sub-paths optimizations, we get a relatively good tour. Figure 6.9 illustrates this process on the small instance solution shown in Fig. 6.3.

Figure 6.10 gives the evolution of the computational time as a function of the number of cities. Figure 6.11 measures the quality of the solutions that can be obtained with these techniques. Interestingly, the greedy nearest neighbor heuristic (Code 4.3) would have provided, in a few 10 years or a few centuries for a billion city instance, a solution deviating by about 22% from the optimum.

6.4.3 Comments

The chief difference between LNS and POPMUSIC is the latter unequivocally defines the stopping criterion and the neighbor solution acceptance. Indeed, POPMUSIC accepts to modify the solution only if we have a strict improvement. For several problems, this framework seems sufficient to obtain good quality solutions, the latter being strongly conditioned by the capacity of the optimization method used. The philosophy of POPMUSIC is to keep a framework as simple as possible. If necessary, the optimization method is improved so that it can better address

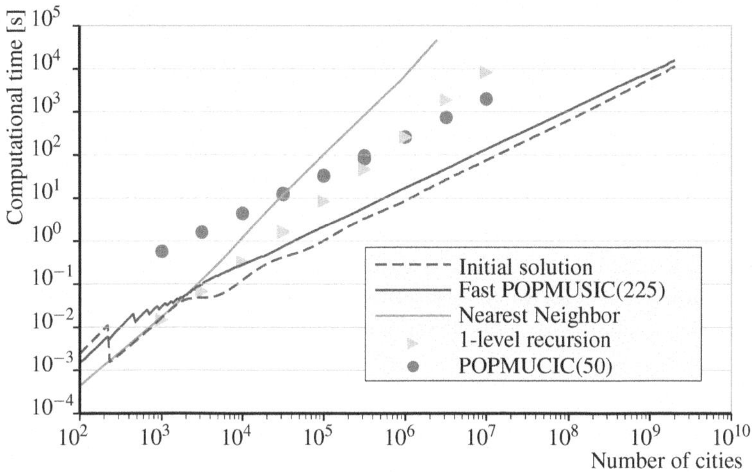

Fig. 6.10 Computational times for building an initial TSP solution with the technique presented in Sect. 6.3.2. Optimizing it with a fast POPMUSIC (sub-paths of 225 cities). Building a solution with the nearest neighbor heuristic. Building a tour with one level recursion method (see Problem 6.4). Optimizing a tour with a standard POPMUSIC (sub-paths of 50 cities). The increase in time for building a solution, in $n \log n$, is higher than that of optimizing it with POPMUSIC. However, the last takes a higher time for an instance with more than 2 billion cities

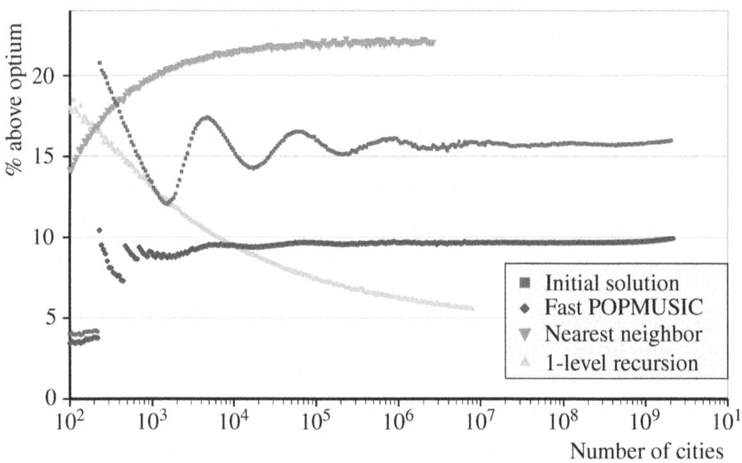

Fig. 6.11 Quality of the solutions obtained with the constructive method presented in Sect. 6.3.2 and once improved with fast POPMUSIC. Quality of the nearest neighbor heuristic and those of a standard POPMUSIC starting from an initial solution obtained with a single recursion level. The problem instances are generated uniformly in the unit square, with toroidal distances (as if the square was folded so that opposite borders are contiguous). For such a distance measure, a statistical approximation of the optimal solution length is known. The fluctuations for the initial solution reflect the recursion levels

larger sub-problems. So, the framework is kept simple, without adding complicated stopping criteria.

Defining parts and their proximity in POPMUSIC is perhaps a more intuitive way than in LNS to formalize a set of constraints that are added to the problem on the basis of an existing solution. These constraints allow using an optimization method that would be inapplicable to the complete instance. The *Corridor Method* [6] takes the problem from the other end: given an optimization method that works well—in their application, dynamic programming—how can we add constraints to the problem so that we can continue to use this optimization method. The components or options of a method are often all interdependent. Choosing one option affects the others. It may explain why actually very similar methods are presented by different names.

Problems

6.1 Dichotomic Search Complexity
By applying the master recurrence theorem (Sect. 5.2.1), determine the algorithmic complexity of searching for an element in a sorted array by means of a dichotomic search.

6.2 POPMUSIC for the Flowshop Sequencing Problem
For implementing a POPMUSIC-based method, how to define a part and a sub-problem for the flowshop sequencing problem? How to take into account the interaction between the sub-problem and parts that should not be optimized?

6.3 Algorithmic Complexity of POPMUSIC
In a POPMUSIC application, the size of the sub-problem is independent of the size of the problem instance. Hence, any sub-problem can be solved in a constant time. Empirical observations, like those presented in Fig. 6.8, show that the number of times a portion is inserted in U is also independent of the instance size. In terms of algorithmic complexity, what are the most complex steps of POPMUSIC?

6.4 Minimizing POPMUSIC Complexity for the TSP
A technique for creating an appropriate TSP tour is as follows: first, a random sample of k cities among n is selected. A good tour on the sample is obtained with a heuristic method. Let us suppose that the complexity of this method is $O(k^a)$, where a is a constant larger than 1. Then, for each of the remaining $n - k$ cities, we find the nearest from the sample. In the partial tour, each remaining city is inserted (in any order) just after the sample city identified as the nearest. Finally, sub-paths of r cities of the tour thus obtained are optimized with POPMUSIC. The value of r is supposed to be in $O(n/k)$. Also, it is supposed that the total number of sub-paths optimized with POPMUSIC is in $O(n)$. The paths are optimized using the same heuristic method as for finding a tour on the sample.

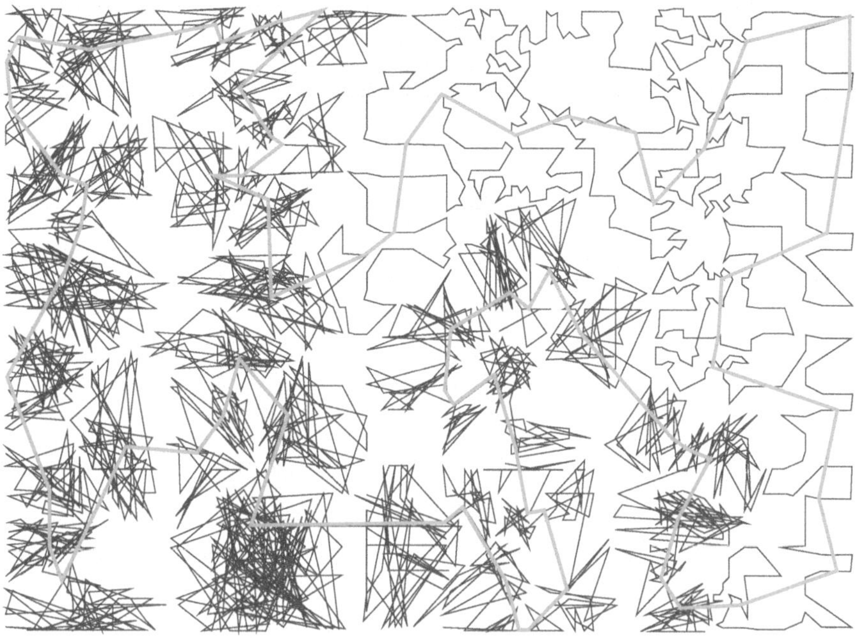

Fig. 6.12 TSP tour partially optimized with POPMUSIC. The initial tour is obtained with a one-level recursive method. The tour on a sample of the cities in bold

Figure 6.12 illustrates the process. The sample size k depends on the number of cities. We suppose that $k = \Theta(n^h)$, where h is to be determined. The sub-paths optimized with POPMUSIC have a number of cities proportional to n/k. Determine the value of $h(a)$ that minimizes the global complexity of this method.

References

1. Adams, J., Balas, E., Zawack, D.: The shifting bottleneck procedure for job shop scheduling. Manag. Sci. **34**(4), 391–401 (1998). https://doi.org/10.1287/mnsc.34.3.391
2. Alvim, A.C.F., Taillard, É.D.: POPMUSIC for the world location routing problem. EURO J. Transp. Log. **2**(3), 231–254 (2013). https://doi.org/10.1007/s13676-013-0024-2
3. Greistorfer, P., Staněk, R., Maniezzo, V.: The magnifying glass heuristic for the generalized quadratic assignment problem. In: Metaheuristic International Conference (MIC'19) Proceedings. Cartagena, Columbia (2019)
4. Sahling, F., Buschkühl, L., Tempelmeier, H., Helber, S.: Solving a multi-level capacitated lot sizing problem with multi-period setup carry-over via a fix-and-optimize heuristic. Comput. Oper. Res. **36**(9), 2546–2553 (2009). https://doi.org/10.1016/j.cor.2008.10.009
5. Shaw, P.: Using constraint programming and local search methods to solve vehicle routing problems. In: International Conference of Principles and Practice of Constraint Programming. 417–431. Springer (1998). https://doi.org/10.1007/3-540-49481-2_30

6. Sniedovich, M., Voß S.: The Corridor method: A dynamic programming inspired metaheuristic. Control and Cybernetics. **35**(3), 551–578 (2006). http://eudml.org/doc/209435
7. Taillard, É.D.: Parallel iterative search methods for vehicle routing problems. Networks. **23**(8), 661–673 (1993). https://doi.org/10.1002/net.3230230804
8. Taillard, É.D.: La programmation à mémoire adaptative et les algorithmes pseudo-gloutons: nouvelles perspectives pour les méta-heuristiques. HDR thesis, Université de Versailles-Saint-Quentin-en-Yvelines (1998)
9. Taillard, É.D.: Heuristic methods for large centroid clustering problems. J. Heuristics. **9**(1), 51–73 (2003). https://doi.org/10.1023/A:1021841728075
10. Taillard, É.D.: A linearithmic heuristic for the travelling salesman problem. Eur. J. Oper. Res. **297**(2), 442–450 (2022). https://doi.org/10.1016/j.ejor.2021.05.034

Part III
Popular Metaheuristics

This part reviews about 20 popular metaheuristics. They are grouped by the core of the basic heuristic principle they exploit. The first group includes those relying exclusively on random components. Second are methods that attempt to learn how to build new solutions. This is followed by methods that learn how to modify solutions with a local search. Finally, there are methods that exploit a population of solutions. Table 1 lists the metaheuristics presented in this part.

The last chapter provides tips for designing new heuristics.

Table 1 Metaheuristics that are addressed in the third part with a brief description of their operating principles

Chapter	Method	Operating principles
7. Randomized methods	Simulated annealing	Biased random local search
	Threshold accepting	
	Great deluge	
	Demon algorithm	
	Noising methods	
	Late acceptance hill climbing	Local search + history
	Variable neighborhood search	Several neighborhoods
	GRASP	Biased random construction + local search
8. Construction learning	MIN-MAX ant system	Construction with learning + local search
	FANT	
	Vocabulary building	Advanced construction learning
9. Local search learning	Taboo search	Local search + memory
	Strategic oscillations	Local search + various memories
10. Population management	Genetic algorithm	Simple population evolution
	Memetic algorithm	Advanced population management + local search
	Scatter search	
	BRKGA	
	Path relinking	
	GRASP-PR	
	Fixed set search	
	Particle swarm	Interactions between continuous solutions
	Electromagnetic method	

Chapter 7
Randomized Methods

By applying the principles presented in the previous chapters, we can build a solution and improve it to find a local optimum. In addition, if the problem is complicated or large, we have seen how to decompose it into sub-problems easier to solve.

What if, using these techniques, we obtain solutions whose quality is not good enough? Let us suppose that we can devote more computational time to finding better solutions. The first option coming to mind is to try to introduce a learning process. That is the subject of subsequent chapters.

A second option—a priori simpler to implement—is to incorporate random components into an "improvement" method where we allow the choice of lower quality solutions than that of departure. Although we no longer have a strict improvement at each iteration, this is a *local search* since such methods are based on locally modifying solutions. Very similar methods are based on this second option: *simulated annealing* (SA), *threshold accepting* (TA), *great deluge*, *demon algorithms*, and the *noising method*. Interestingly, the latter framework can be seen as a generalization of the previous methods. The *late acceptance hill climbing* shares similarities with these methods but incorporates a self-parameter tuning. The *variable neighborhood search* (VNS) alternates *intensification* and *diversification* phases. It improves a solution with a basic neighborhood and degrades it by randomly selecting moves in other neighborhoods.

A third option is to repeat constructions with random choices, possibly followed by local improvements. This is the way followed by the *greedy randomized adaptive search procedure* (GRASP).

É. D. Taillard, *Design of Heuristic Algorithms for Hard Optimization*, Graduate Texts in Operations Research, https://doi.org/10.1007/978-3-031-13714-3_7

7.1 Simulated Annealing

The simulated annealing method is one of the first local search techniques that does not strictly improve the quality of the solution at each iteration. This method is inspired by a process of physics, the annealing, which minimizes the internal energy of the molecules of a material. Some materials, like metals, have their internal structure modified, depending on the temperature to which they are heated. By rapidly cooling the material, the molecules do not have time to arrange to achieve the usual structure at low temperatures but forms grains which are small crystals whose orientation is different for each grain. This is the quenching process, which is used in particular to harden some steels.

On the contrary, if the cooling is slow, the molecules manage to form crystals much larger, corresponding to their minimum energy state. By repeating the method, one can further increase the size of the crystals or even obtain a monocrystal. This is the *annealing* process. These two processes are illustrated in Fig. 7.1.

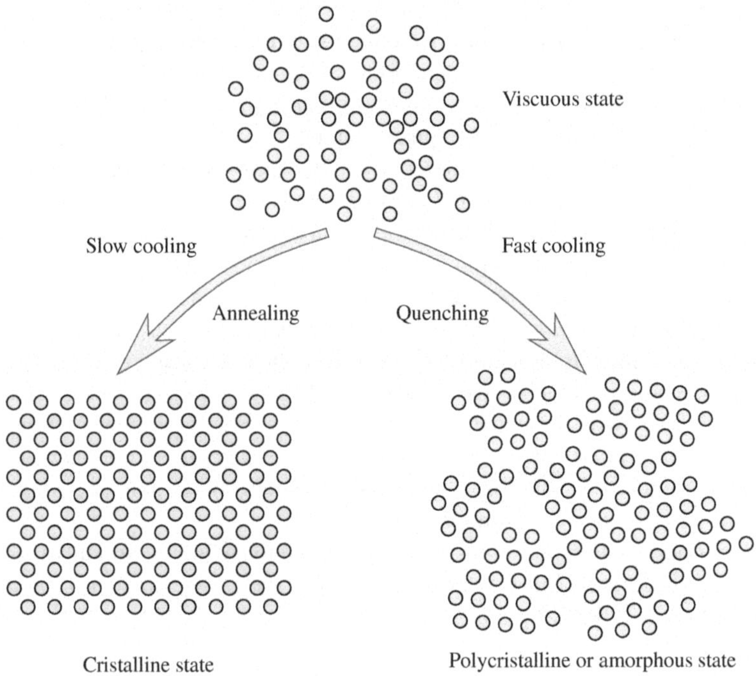

Fig. 7.1 Annealing and quenching processes. The material is carried to such a temperature that its molecules have enough energy to move. By cooling it slowly, the molecules have time to reach a crystalline state, minimizing their energy. This is the annealing process. If the cooling is faster, this is the quenching process, and disordered crystals are formed. Under certain conditions, a very fast cooling does not allow time to for crystals to form and the material remains in an amorphous state

Černý [1] and Kirkpatrick et al. [8] independently had the idea to simulate this process in combinatorial optimization, making the analogy between the objective function to minimize and the energy of the molecules. At high temperatures, a molecule has enough energy to fill a gap in the crystal lattice or change its configuration. However, at low temperatures, it has a significantly lower probability of doing so. Translating this in terms of combinatorial optimization means changing a solution locally and randomly and accepting its degradation with a certain probability. The latter must be low if the degradation is significant.

Expressed in terms of local search, this corresponds to generating a random move $m \in M$ and calculating the cost difference Δ between the initial and modified solution: $\Delta = f(s \oplus m) - f(s)$. If $\Delta < 0$, the move m improves s, and it is accepted. The new solution becomes $s \oplus m$. Else, the move m can eventually be accepted, with a probability proportional to $e^{-\Delta/T}$, where T is a parameter simulating the temperature. At each step, the temperature T is diminished. Several formulas have been proposed to adjust the temperature. Among the most frequently encountered, $T \leftarrow \alpha \cdot T$ and $T \leftarrow \frac{T}{1+\alpha T}$, where $0 < \alpha < 1$, is the parameter adjusting the decreasing speed of the temperature. The method comprises at least two other parameters: T_{init} and T_{end} the initial and finishing temperatures. Algorithm 7.1 provides the framework for basic simulated annealing.

Algorithm 7.1: Elementary simulated annealing. Countless variants of algorithms based on this framework have been proposed. Practical implementations do not return the solution s of the last iteration but the best solution found throughout the search

Input: Initial solution s; fitness function f to minimize; neighborhood structure M, parameters T_{init}, $T_{end} < T_{init}$ and $0 < \alpha < 1$

Result: Modified solution s

$T \leftarrow T_{init}$
while $T > t_{end}$ **do**
  Randomly generate $m \in M$
  $\Delta = f(s \oplus m) - f(s)$
  Randomly generate $0 < u < 1$
  if $\Delta < 0$ *or* $e^{-\Delta/T} > u$ **then** m is accepted
    $s \leftarrow s \oplus m$
  $T \leftarrow \alpha \cdot T$

This framework is generally modified. First, the parameters defining the initial and final temperatures provide a very different effect according to the fitness function measure unit. Indeed, if f measures the length of the TSP tour, these parameters should be adapted depending on whether the unit is meters or kilometers. To make the algorithm more robust, we do not invite the user to directly provide temperatures. For instance, the user specifies degrading moves acceptance rates τ_{init} et τ_{end}, which is much more intuitive. The temperatures are then calculated

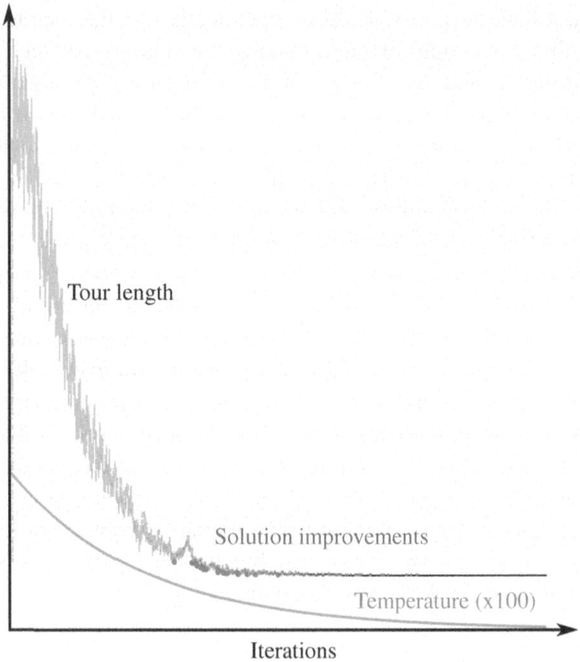

Fig. 7.2 Changes in TSP tour length of and temperature evolution during simulated annealing

automatically according to these rates. A random walk performing a few hundred or a few thousand steps can record statistics on the average Δ values.

Frequently, the temperature is not decreased at each iteration (Line 8). Another parameter is introduced, defining the number of iterations performed with a given temperature.

Figure 7.2 illustrates the evolution of the tour length for a TSP with 225 cities. Code 7.1 was executed with an initial temperature of $5 \cdot d_{max}/n$, a final temperature of $20 \cdot d_{max}/n^2$ and $\alpha = 0.99$, where d_{max} is the largest distance between two cities. The algorithm was provided a relatively good initial tour. There is a significant deterioration of the latter during the first iterations at high temperatures. This degradation is necessary to alter the structure of the starting solution to discover better solutions. About half of the iterations are carried out unnecessarily in this run, as the value of the best solution found no longer evolves.

Code 7.1 implements a very basic simulated annealing for the TSP. It is based on the 2-opt neighborhood structure. Algorithm 7.1 is adapted to decrease the temperature only every n^2 iterations and not at each iteration. Thus, a value of α between 0.8 and 0.99 produces satisfactory results, regardless of the instance size. Finally, the user must provide the absolute value of the initial and final temperatures. As mentioned above, $T_{init} = 5 \cdot d_{max}/n$ and $T_{end} = 20 \cdot d_{max}/n^2$ are values that can be suitable for starting a parameter tuning.

Code 7.1 tsp_SA.py A basic simulated annealing implementation for the TSP

```
import math
from random_generators import unif, rando                       # Listing 12.1

######### Basic Simulated Annealing for the TSP, based on 2-opt moves
def tsp_SA(d,                                                  # Distance matrix
           tour,                                                    # TSP tour
           length,                                               # Tour length
           initial_temperature,                                # SA parameters
           final_temperature,
           alpha):

    n = len(tour)
    best_length = length
    best_tour = tour
    T = initial_temperature
    iteration = 0
    while T > final_temperature:
        i = unif(0, n-1)                  # First city of a move randomly chosen
        j = (i + unif(2, n-2))%n        # Second city is unif successors further
        if j < i:
          i, j = j, i                              # j must be further on the tour
        delta =  d[tour[i]][tour[j]]      + d[tour[i+1]][tour[(j+1) % n]]\
                -d[tour[i]][tour[i + 1]] - d[tour[j]][tour[(j + 1) % n]]
        if delta < 0 or math.exp(-delta / T) > rando():
            length = length + delta                               # Move accepted
            for k in range((j - i) // 2):     # Reverse sub-path between i and j
                tour[k + i + 1], tour[j - k] = tour[j - k], tour[k + i + 1]

        # is there an improvement?
        if best_length > length:
            best_length = length
            best_tour = tour
            print('SA {:d} {:d}'.format(iteration, length))
        iteration += 1
        if iteration % (n * n) == 0:
            T *= alpha                                      # Decrease temperature
    return best_tour, best_length
```

7.2 Threshold Accepting

Threshold accepting, proposed by Dueck and Scheuer [4], is a pure local search. It only moves from a solution to one of its neighbors. Like simulated annealing, demon, and great deluge algorithms, the move are randomly chosen but are not systematically applied to the current solution. In the case of an improving move, the neighbor solution is accepted. If the move deteriorates the solution, it is only accepted if the deterioration is lower than a given threshold. The latter is gradually decreased to reach zero, so that the method stops in a local optimum. Algorithm 7.2 provides the threshold accepting framework.

Algorithm 7.2: Threshold accepting. The values of the thresholds $\tau_1, \ldots \tau_T$ are not necessarily explicitly provided but calculated, for instance, by providing only the initial threshold and multiplying it by another parameter, α, at each round of R iterations

Input: Initial solution s; fitness function f to minimize; neighborhood structure M, parameters T, R, $\tau_1, \ldots \tau_T$
Result: Solution s^*

$s^* \leftarrow s$
for *t from* 1 *to* T **do**
  for *R iterations* **do**
    Randomly generate $m \in M$
    if $f(s \oplus m) - f(s) < \tau_t$ **then** the move m is accepted
      $s \leftarrow s \oplus m$
      if $f(s) < f(s^*)$ **then**
        $s^* \leftarrow s$

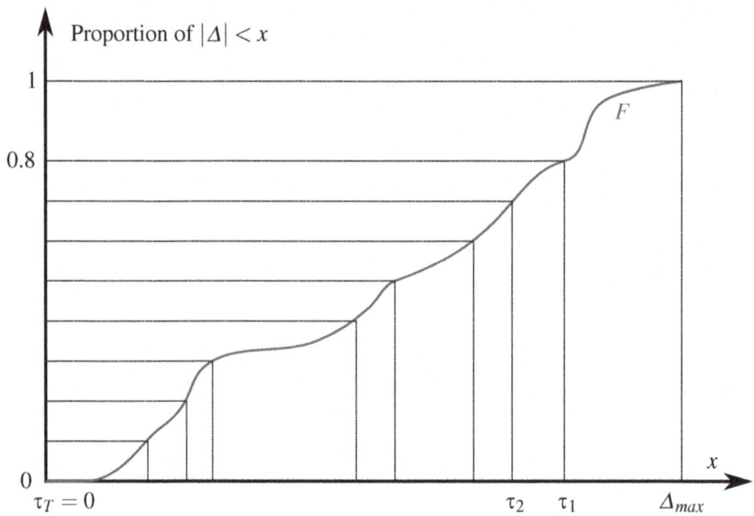

Fig. 7.3 Technique for determining the thresholds $\tau_1, \ldots, \tau_T$. The empirical distribution function F is obtained by performing a number of random moves and recording their amplitude in absolute value

Gilli et al. [6] proposed an automated method for setting the thresholds. First, random moves are performed, recording the fitness difference of neighbor solutions. This allows determining the empirical distribution function F of the amplitude of the moves. Then, the T thresholds are fixed using the inverse of this function (see Fig. 7.3): $\tau_t = F^{-1}(0.8 \cdot (T - t)/T)$. Put differently, the first threshold τ_1 accepts about 80% of the degrading moves, while the last, $\tau_T = 0$, only allows improvements.

7.3 Great Deluge Algorithm

The great deluge algorithm, proposed by Dueck [3], has similarities with the previous one. However, the absolute value of the fitness function limits the search progression instead of the amplitude of the moves. The name of this method comes from the legend that, as a result of incessant rain, all terrestrial beings eventually drowned, except for those on Noah's Ark. The animals on the ground panicked and ran in all directions, everywhere except where there was water.

The analogy with a maximization process is made by considering random moves that are accepted as long as they do not lead to a solution whose quality is less than a threshold L. The last is the water level which increases by a value of P at each iteration. This parameter simulates the rain strength. The process stops when the value of the current solution is less than L. Algorithm 7.3 provides the framework of this method. Its operating principle is illustrated in Fig. 7.4.

Algorithm 7.3: Great deluge algorithm. The algorithm can adapt to minimization problems. Hence, simulating the behavior of fish when the water level drops!

Input: Solution s, fitness function f to maximize, neighborhood structure M, parameters L and P

Result: s^*

$s^* \leftarrow s$
while $f(s) > L$ **do**
  Generate a random move $m \in M$
  if $f(s \oplus m) > L$ **then**
    $s \leftarrow s \oplus m$
    if $f(s) < f(s^*)$ **then**
      $s^* \leftarrow s$
    $L \leftarrow L + P$

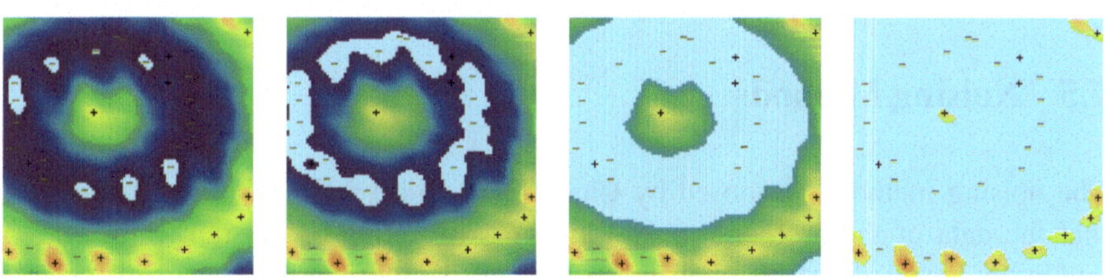

(a) (b) (c) (d)

Fig. 7.4 Illustration of the great deluge algorithm. State of the landscape at the beginning of the rain (**a**). The water level increases, (**b**), (**c**), until only the highest peaks emerge (**d**)

7.4 Demon Algorithm

The demon algorithm simulates the behavior of a compulsive gambler who always bets larger amounts of money. The devil of playing pushes futilely to spend money if the earnings exceed a certain threshold, D_{max}. Once this threshold is reached, the gambler continues to play. The gambler enters the casino with an sum of D in pocket and stops playing when exhausted, after I_{max} bets. This last parameter is involved in many iterative local searches for directly adjusting the computational effort. Translated in terms of local search, a bet is to allow a degrading move. But the loss cannot exceed the available budget. If the move improves the solution, the budget is increased accordingly, up to the maximum threshold. Algorithm 7.4 provides the framework of this method.

Algorithm 7.4: Demon algorithm. This algorithm is relatively simple to implement, but, as for threshold accepting, its parameters must be adjusted according to the numerical value of the data

Input: Solution s, fitness function f to minimize, neighborhood structure M, parameters I_{max}, D, D_{max}

Result: s^*

$s^* \leftarrow s$

for I_{max} *iterations* **do**
- Randomly generate a move $m \in M$
- $\Delta = f(s \oplus m) - f(s)$
- **if** $\Delta \leqslant D$ **then**
  - $D \leftarrow D - \Delta$
  - **if** $D > D_{max}$ **then**
    - $D \leftarrow D_{max}$
  - $s \leftarrow s \oplus m$
  - **if** $f(s^*) > f(s)$ **then**
    - $s^* \leftarrow s$

7.5 Noising Methods

The noising methods, proposed by Charon and Hudry [2], make the assumption that the data of the problem are not known with infinite precision. Under these conditions, even if the problem is convex and simple, an improvement method can be trapped in local optima resulting from artifacts. To make the improvement method more robust, a random noise is added, either to the data or to the move evaluation. For instance, the coordinates of a Euclidean TSP can be slightly changed. Taking these stochastic values into account, the resulting fitness function has no local optima. The lasts are somehow erased by the random noise. The framework of noising methods is provided by Algorithm 7.5.

Algorithm 7.5: Noising method. At each move evaluation, random noise is generated according to the probability distribution $noise(i)$, whose variance generally decreases with i

Input: Solution s, fitness function f to minimize, neighborhood structure M, parameters $I_{max}, noise(i)$

Result: s^*

$s^* \leftarrow s$

forall $i \in 1 \dots I_{max}$ **do**

  Randomly generate a move $m \in M$

  if $f(s \oplus m) + noise(i) < f(s)$ **then**

    $s \leftarrow s \oplus m$

    if $f(s) < f(s^*)$ **then**

      $s^* \leftarrow s$

A parameter of the method is a probability distribution, set up with the iteration number. Each time a solution is evaluated, a random noise occurrence is generated. Generally, its expectation is zero, and its variance decreases with the iteration number (see Fig. 7.5).

At the end of the algorithm, the method gets closer and closer to an improvement method. The variance of the noise function must naturally depend on the numerical data of the problem. To achieve this goal, one can incorporate the evaluation of

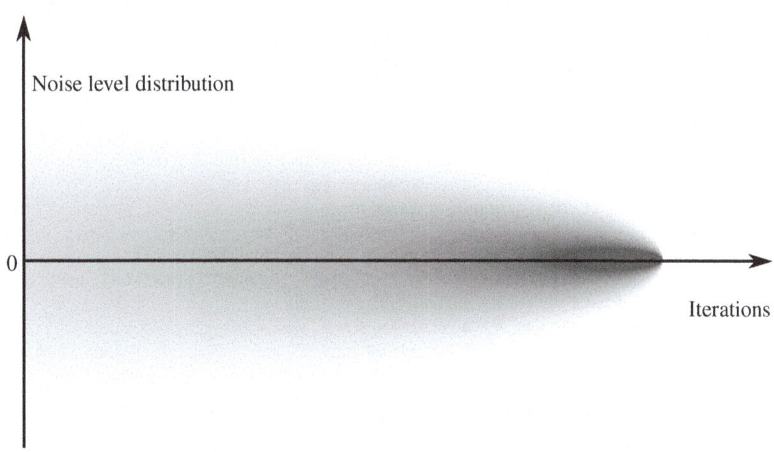

Fig. 7.5 Usual example of the evolution of the random noise distribution as a function of the number of iterations performed by a noising method. The color density represents the probability density

the objective function in the probability distribution. The choice of the latter can transform a noising method into simulated annealing, threshold accepting or other techniques described above. Code 7.2 implements a kind of noising method for the TSP.

Code 7.2 tsp_noising.py Implantation of a noising method for the TSP. The noise distribution is uniform and decreases exponentially with the number of iterations performed

```
from build_2opt_data_structure import *                          # Listing  5.3
from random_generators import unif, rando                        # Listing 12.1
from tsp_utilities import *                                      # Listing 12.2
from math import *

######### Noising method for the TSP
def tsp_noising(d,                                             # Distance matrix
                tour,                                                 # TSP tour
                length,                                            # Tour length
                initial_noise,                                      # Parameters
                final_noise,
                alpha):

    n = len(tour)
    t = build_2opt_data_structure(tour)
    current_noise = initial_noise
    best_length = length
    iteration = 0
    while current_noise > final_noise:
        i = unif(0, n-1)                     # First city of a move randomly chosen
        last_i = i
        while t[t[i]]>>1 != last_i and t[i]>>1 != last_i:
            j = t[t[i]]
            while j>>1 != last_i and (t[j]>>1 != last_i or i>>1 != last_i):
                delta = d[i >> 1][j >> 1]     + d[t[i] >> 1][t[j] >> 1] \
                       -d[i >> 1][t[i] >> 1] - d[j >> 1][t[j] >> 1]
                if delta + current_noise * log(rando()) < 0:     # SA criterion
                    length = length + delta                     # Move accepted
                    best_i, best_j = t[i], t[j]
                    t[i], t[j] = j ^ 1, i ^ 1 # New successors and predecessors
                    t[best_i ^ 1], t[best_j ^ 1] = best_j, best_i

                    i = t[i]      # Avoid reversing immediately a degrading move
                    j = t[i]

                    # is there an improvement?
                    if best_length > length:
                        best_length = length
                        tour = tsp_2opt_data_structure_to_tour(t)
                        print('Noising {:d} {:d}'.format(iteration, length))
                iteration += 1
                if iteration % (n * n) == 0:
                    current_noise *= alpha                      # Decrease noise
                j = t[j]                                                # Next j
            i = t[i]                                                    # Next i
    return tour, best_length
```

To evaluate and perform a move in constant time, the data structure introduced in Section 5.1.3.3 has been used. However, this data structure does not allow evaluating a random move (i, j) in constant time. Indeed, performing the tour in a given direction, given a city i and its succeeding one s_i, we cannot immediately identify the city s_j that succeeds j.

The artifice used in this code is to systematically scan the whole neighborhood instead of randomly drawing the move. The noise added to the evaluation of a move is such that the acceptance criterion is identical to that of a simulated annealing.

7.6 Late Acceptance Hill Climbing

Another technique is similar to the methods of simulated annealing, threshold accepting, and great deluge. The core idea is to differ the acceptance criterion of a neighbor solution. Instead of comparing the value of the latter with that of the current solution, it is compared to that obtained h iterations before. Thus, a neighbor solution is accepted either if it is at least as good as the current solution, or if it is better than the solution visited h iterations previously. The derived method is called *Late Acceptance Hill Climbing (LAHC)*.

"Hill climbing" refers to an improvement method seeking to maximize an objective. Naturally, a descent method is obtained by changing the acceptance criterion of a neighbor solution. LAHC implementation requires storing a list L of the h values of the previous solution visited. This strategy allows a self-calibration of the acceptance criterion of a worse solution. Unlike the methods viewed above, LAHC is insensitive to the order of magnitude of the fitness values.

The framework of LAHC, given by Algorithm 7.6, does not specify a stopping criterion. A possibility is to set the probability p of being in a local optimum relative to the neighborhood M. The stopping criterion corresponding to this probability is to have performed $|M|(\ln |M| - \ln(1-p))$ iterations without improving the current solution.

7.7 Variable Neighborhood Search

Variable neighborhood search (VNS) [7] implements an idea called *strategic oscillations*. The search alternates intensification and diversification phases. The chief idea of this method is to rely on several neighborhoods $M_1 \dots M_p$. A first neighborhood, M_1, is exploited as usual to find local optima. This is one of the most elementary ways to intensify the search. The other neighborhoods allow escaping from these optima by performing random moves. The latter ones increasingly destroy the local optima structure. Performing random moves is also one of the simplest ways to diversify the search for exploring other portions of the solution space. The framework of a basic VNS is provided by Algorithm 7.7.

A very limited number of neighborhoods are generally used. In this case, the method is repeated several times (see Problem 7.4). Several variants of this

Algorithm 7.6: Late acceptance hill climbing

Input: Solution s, fitness function f to minimize, neighborhood structure M, parameters h, stopping criterion
Result: Improved solution s

for $k \in 0 \dots h-1$ **do**
  $L_k \leftarrow f(s)$
$i \leftarrow 0$
repeat
  Randomly select a move $m \in M$
  if $f(s \oplus m) < L_i$ *or* $f(s \oplus m) \leqslant f(s)$ **then**
    $s \leftarrow s \oplus m$
  if $f(s) < L_i$ **then**
    $L_i \leftarrow f(s)$
  $i \leftarrow (i+1) \mod h$
until *The stopping criterion is satisfied*

Algorithm 7.7: Variable neighborhood search. When a limited number p of neighborhoods are available, the algorithm is repeated several times

Input: Solution s, fitness function f to minimize, neighborhood structures $M_1 \dots M_p$
Result: s^*

$s^* \leftarrow s$
$k \leftarrow 1$
while $k \leqslant p$ **do**
  Randomly generate a move $m \in M_k$
  $s \leftarrow s \oplus m$
  Find the local optimum s' associated with s in neighborhood M_1
  if $f(s') < f(s^*)$ **then**
    $s^* \leftarrow s'$
    $k \leftarrow 1$
  else
    $s \leftarrow s^*$
    $k \leftarrow k+1$

framework have been proposed: VNS descent, VNS decomposition, skewed VNS, and reduced VNS.

Code 7.3 provides a very simple VNS implementation for the TSP.

Code 7.3 tsp_VNS.py VNS implementation for the TSP. The neighborhood M_k consists in swapping two cities k times. The repair method is an ejection chain. In addition to its extreme simplicity, this implementation requires no parameters

```python
from random_generators import unif                          # Listing 12.1
from tsp_utilities import tsp_length                        # Listing 12.2
from tsp_LK import tsp_LK                                   # Listing 12.3

######### Variable Neighborhood Search for the TSP
def tsp_VNS(d,                                              # Distance matrix
            best_tour,                                      # TSP tour
            best_length):

    n = len(best_tour)
    iteration, k = 0, 1
    while k < n:
        tour = best_tour[:]

        for _ in range(k):                                  # Perturbate solution
            u = unif(0, n - 1)
            v = unif(0, n - 1)
            tour[u], tour[v] = tour[v], tour[u]
        length = tsp_length(d, tour)
        tour, length = tsp_LK(d, tour, length)
        iteration += 1
        if length < best_length:
            best_tour = tour[:]
            best_length = length
            print('VNS {:d}\t {:d}\t {:d}'
                  .format(iteration, k, length))
            k = 1
        else:
            k += 1
    return best_tour, best_length
```

7.8 GRASP

The *greedy randomized adaptive search procedure (GRASP)* was proposed by Feo and Resende [5]. It repeatedly improves, with a local search, a solution obtained with a greedy constructive method. The last incorporates a random component so that it produces various solutions. This method comprises two parameters, I_{max}, the number of repetitions of the outer loop of the algorithm and α, to adjust the degree of randomization. The framework of the method is provided by Algorithm 7.8.

Code 7.4 implements the construction of a solution using GRASP. In practice, this code is repeated several times, and the best solutions produced are retained.

Algorithm 7.8: GRASP. The programmer must initially design a method, *local_search*, for improving a solution. The user must provide two parameters, I_{max} which sets the computational effort, and α which sets the random choice level. The difference with the greedy constructive Algorithm 4.2 are highlighted

Input: Set E of elements constituting a solution; incremental cost function $c(s,e)$; fitness function f to minimize, parameters I_{max} and $0 \leqslant \alpha \leqslant 1$, improvement method *local_search*

Result: Complete solution s^*

$f^* \leftarrow \infty$
for *I_{max} iterations* **do**
  Initialize s to a trivial partial solution
  $R \leftarrow E$ // Elements that can be added to s
  while $R \neq \varnothing$ **do**
    Find $c_{min} = \min_{e \in R} c(s,e)$ and $c_{max} = \max_{e \in R} c(s,e)$
    Choose randomly, uniformly $e' \in R$ such that $c_{min} \leqslant c(s,e') \leqslant c_{min} + \alpha(c_{max} - c_{min})$
    $s \leftarrow s \cup e'$ // Include e' in the partial solution s
    Remove from R the elements that cannot be added any more to s
  $s' \leftarrow local\_search(s)$ // Find the local optimum associated with s
  if $f^* > f(s')$ **then**
    $f^* \leftarrow f(s')$
    $s^* \leftarrow s'$

Code 7.4 tsp_GRASP.py GRASP Implementation for the TSP. The randomized greedy construction is based on the nearest neighbor criterion. The improvement procedure is the ejection chain Code 12.3

```
from random_generators import rand_permutation                    # Listing 12.1
from tsp_utilities import tsp_length                              # Listing 12.2
from tsp_LK import tsp_LK                                         # Listing 12.3

######### Procedure for producing a TSP tour using GRASP principles
def tsp_GRASP(d,                                                  # Distance matrix
              alpha):

    n = len(d[0])
    tour = rand_permutation(n)
    for i in range(n - 1):
        # determine c_min and c_max incremental costs
        c_min, c_max = float('inf'), float('-inf')
        for j in range(i + 1, n):
            if c_min > d[tour[i]][tour[j]]:
                c_min = d[tour[i]][tour[j]]
            if c_max < d[tour[i]][tour[j]]:
                c_max = d[tour[i]][tour[j]]

        next = i+1          # Find the next city to insert, based on lower cost
        while d[tour[i]][tour[next]] > c_min + alpha * (c_max - c_min):
            next += 1
        tour[i + 1], tour[next] = tour[next], tour[i + 1]

    length = tsp_length(d, tour)
    tour, length = tsp_LK(d, tour, length)
    return tour, length
```

Setting parameter $\alpha = 0$ leads to a purely greedy constructive method. Unless many elements have the same incremental cost, the repetition of constructions does not make sense. Setting $\alpha = 1$ leads to a purely random construction. The method represents then an iterative local search starting with random solutions. This technique is often used because it produces better solutions than a single local search run and requires negligible coding effort. To benefit from the advantages of the GRASP method, it is necessary to tune the α parameter. Usually, it will produce its full potential for values close to 0. It should additionally be noted that the initialization of the partial solution may include a random component. For the TSP, it can be the departure city. The incremental cost function may correspond to the nearest neighbor criterion.

Problems

7.1 SA Duration

How many iterations does a simulated annealing run if it starts with an initial temperature T_0 and ends at temperature T_f, knowing that the temperature is multiplied by α at each iteration?

7.2 Tuning GRASP

Try to tune the α parameter of the GRASP code. Take the TSPLIB problem instance `tsp225`. Are good values depending on the number of iterations I_{max} the method performs?

7.3 VNS with a Single Neighborhood

VNS requires to have p different neighborhoods and to use a particular neighborhood M_1 to find local optima. If we only have M_1, how can we build the other neighborhoods?

7.4 Record to Record

In `tsp_VNS`, n different neighborhoods are used, leading to a method without parameters. Could a more efficient algorithm be obtained by limiting the number of neighborhoods and repeating the search several times? How many neighborhoods and how many times should we repeat the search?

References

1. Černý, V.: Thermodinamical approach to the traveling salesman problem: an efficient simulation algorithm. J. Optim. Theory Appl. **45**(1), 41–51 (1985). https://doi.org/10.1007/BF00940812
2. Charon, I., Hudry, O.: The Noising method: a new method for combinatorial optimization. Oper. Res. Lett. **14**(3), 133–137 (1993). https://doi.org/10.1016/0167-6377(93)90023-A

3. Dueck, G.: New optimization heuristics: the great deluge algorithm and the record-to-record travel. J. Comput. Phys. **104**(1), 86–92 (1993). https://doi.org/10.1006/jcph.1993.1010
4. Dueck, G., Scheuer, T.: Threshold accepting: a general purpose optimization algorithm appearing superior to simulated annealing. J. Comput. Phys. **90**(1), 161–175(1990). https://doi.org/10.1016/0021-9991(90)90201-B
5. Feo, T.A., Resende, M.G.C.: Greedy randomized adaptive search procedure. J. Global Optim. **6**, 109–133 (1995). https://doi.org/10.1007/BF01096763
6. Gilli, M., Këllezi, E., Hysi, H.: A data-driven optimization heuristic for downside risk minimization. J. Risk. **8**(3), 1–19 (2006)
7. Hansen, P., Mladenović, N.: An introduction to variable neighborhood search. In: Voß S. , Martello, S., Osman, I.H., Roucairol, C. (eds.) Meta-Heuristics: Advances and Trends in Local Search Paradigms for Optimization. pp. 422–458. Kluwer, Dordrecht (1999). https://doi.org/10.1007/978-1-4615-5775-3_30
8. Kirkpatrick, S., Gelatt, C.D., Vecchi, M.P.: Optimization by simulated annealing. Science. **220** (4598), 671–680 (1983). https://doi.org/10.1126/science.220.4598.671

Chapter 8
Construction Learning

After having studied the four basic principles—modeling, decomposition, construction, and improvement—this chapter introduces the fifth principle of metaheuristics: learning mechanisms. The algorithms seen in the previous chapter rely solely on chance to try to obtain better solutions than would be provided by greedy constructive methods or local searches. This is probably not very satisfactory from the intellectual point of view. Without solely relying upon chance, this chapter studies how to implement learning techniques to build new solutions. Learning processes require three ingredients:

- Repeating experiences and analysing successes and failures: we only learn by making mistakes!
- Memorizing what has been made.
- Forgetting the details. This gives the ability to generalize when in a similar but different situation.

8.1 Artificial Ants

The artificial ant technique provides simple mechanisms to implement these learning ingredients in the context of constructing new solutions.

The social behavior of some animals has always fascinated, especially when a population comes to realizations completely out of reach of an isolated individual. This is the case with bees, termites, or ants: although each individual follows an extremely simple behavior, a colony is able to build complex nests or efficiently supply its population with food.

É. D. Taillard, *Design of Heuristic Algorithms for Hard Optimization*, Graduate Texts in Operations Research, https://doi.org/10.1007/978-3-031-13714-3_8

8.1.1 Real Ant Behavior

Following the work of Deneubourg et al. [2] who described the almost algorithmic behavior of ants, researchers had the idea of simulating this behavior to solve difficult problems.

The typical behavior of an ant is illustrated in Fig. 8.1 with an experience made with a real colony that has been isolated. The latter can only look for food by going out from a single orifice. The last is connected to a tube separated into two branches joining further. The left branch is shorter than the one on the right. As ants initially have no information on this fact, the ants equally distribute in both branches (Fig. 8.1a).

While exploring, each ant drops a chemical substance that it is apt to detect with its antennas, which will assist it when returning to the anthill. Such a chemical substance carrying information is called pheromones. On the way back, an ant deposits a quantity of pheromones depending on the quality of the food source. Naturally, an ant that has discovered a short path is able to return earlier than that which used the bad branch.

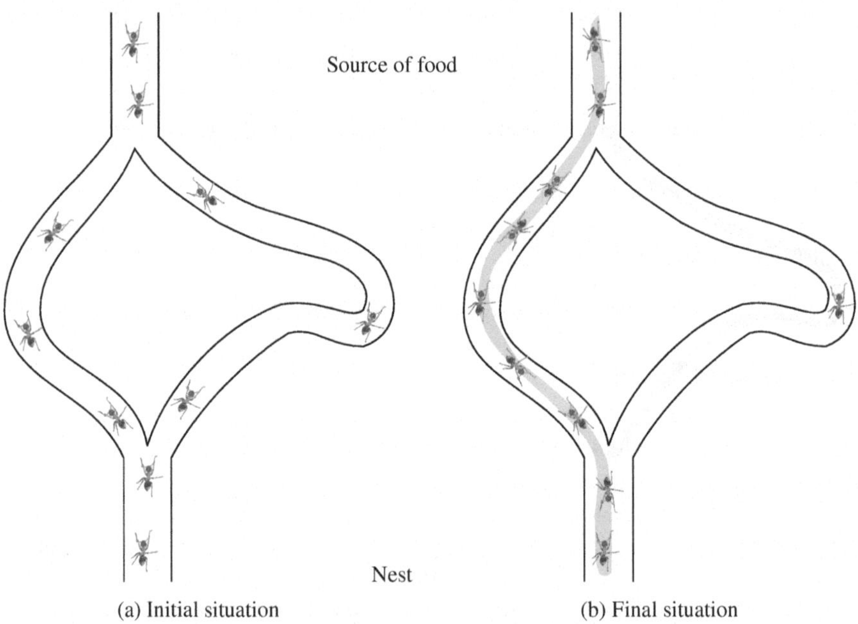

Fig. 8.1 Behavior of an ant colony separated from a food source by a path that is divided. Initially, ants are evenly distributed in both branches (**a**). The ants having selected the shortest path arrive earlier at the food source. Therefore, they faster lay additional pheromones on the way back. The quantity of pheromones deposited on the shortest path grows faster. After a while, virtually all ants will use the shortest branch (**b**)

Therefore, the quantity of pheromones deposited on the shortest path grows faster. Consequently, a new arriving ant has information on the way to take and bias its choice in favour of the shortest branch. After a while, it is observed that virtually all ants use the shortest branch (Fig. 8.1b). Thus, the colony collectively determines an optimal path, while each individual sees no further than the tip of its antennas.

8.1.2 Transcription of Ant Behavior to Optimization

If an ant colony manages to optimize the length of a path, even in a dynamic context, we should be able to transcribe the behavior of each individual in a simple process for optimizing intractable problems. This transcript may be obtained as follows:

- An ant represents a process performing a procedure that constructs a solution with a random component. Many of these processes may run in parallel.
- Pheromone trails are τ_e values associated with each element e constituting a solution.
- Traces play the role of a collective memory. After constructing a solution, the values of the elements constituting the latter will be increased by a quantity depending on the solution quality.
- The oblivion phenomenon is simulated by the evaporation of pheromone trails over time.

Next is to clarify how these components can be put in place. The construction process can use a randomized construction technique, almost similar to the GRASP method. However, the random component must be biased not only by the incremental cost function $c(s, e)$, which represents the *a priori* interest of including element e in the partial solution, but also by the value τ_e which is the *a posteriori* interest of this element. The last is solely known after having constructed a multitude of solutions.

The marriage of these two forms of interest is achieved by selecting the next item e to include in the partial solution s with a probability proportional to $\tau_e^\alpha \cdot c(s, e)^\beta$, where $\alpha > 0$ and $\beta < 0$ are two parameters balancing the respective importance accorded to memory and incremental cost. The update of artificial pheromones is performed in two steps, each requiring a parameter. First, the evaporation of pheromones is simulated by multiplying all the values by $1 - \rho$, where $0 \leqslant \rho \leqslant 1$ represents the evaporation rate. Then, each element e constituting a newly constructed solution has its τ_e value increased by a quantity $1/f(s)$, where $f(s)$ is the solution cost, which is assumed to be minimized and greater than zero.

8.1.3 MAX-MIN Ant System

The first artificial ant colony applications contained only the components described above. The trail update is a positive feedback process. There is a bifurcation point between a completely random process (learning-free) and an almost deterministic one, repeatedly constructing the same solution (too fast learning). Therefore, it is difficult to tune a progressive learning process with the three parameters α, β and ρ.

To remedy this, Stützle and Hoos [5] suggested limiting the trails between two values τ_{min} and τ_{max}. Hence, selecting an element is bounded between a minimum and a maximum probability. This avoids elements possessing an extremely high trail value, implying that all solutions would contain these elements. This leads to the MAX-MIN ant system, which proved much more effective than many other previously proposed frameworks. It is given in Algorithm 8.1.

Algorithm 8.1: MAX-MIN ant system framework

Input: Set E of elements constituting a solution; incremental cost function $c(s,e) > 0$; fitness function f to minimize, parameters $I_{max}, m, \alpha, \beta, \tau_{min}, \tau_{max}, \rho$ and improvement method $a(\cdot)$

Result: Solution s^*

$f^* \leftarrow \infty$
for $\forall e \in E$ **do**
  $\tau_e \leftarrow \tau_{max}$
for I_{max} *iterations* **do**
  for $k = 1 \dots m$ **do**
    Initialize s as a trivial, partial solution
    $R \leftarrow E$ // Elements that can be added to s
    while $R \neq \varnothing$ **do** Build a new solution
      Randomly choose $e \in R$ with a probability proportional to $\tau_e^\alpha \cdot c(s,e)^\beta$ // Ant colony formula
      $s \leftarrow s \cup e$
      From R, remove the elements that cannot be added any more to s
    $s_k \leftarrow a(s)$ // Find the local optimum s_k associated with s
    if $f^* > f(s_k)$ **then** Update the best solution found
      $f^* \leftarrow f(s_k)$
      $s^* \leftarrow s_k$
  for $\forall e \in E$ **do** Pheromone trail evaporation
    $\tau_e \leftarrow (1-\rho) \cdot \tau_e$
  $s_b \leftarrow$ best solution from $\{s_1, \dots, s_m\}$
  for $\forall e \in s_b$ **do** Update trail, maintaining it between the bounds
    $\tau_e \leftarrow max(\tau_{min}, min(\tau_{max}, \tau_e + 1/f(s_b)))$

This framework comprises an improvement method. Indeed, implementations of "pure" artificial ants colonies, based solely on building solutions, have proven inefficient and difficult to tune. There may be exceptions, especially for the treatment of highly dynamic problems where an optimal situation at a given time is no longer optimum at another one.

Algorithm 8.1 has a theoretical advantage: it can be proved that if the number of iterations $I_{max} \rightarrow \infty$ and if $\tau_{min} > 0$, then it finds a globally optimal solution with probability tending to one. The demonstration is based on the fact that $\tau_{min} > 0$ implies that the probability of building a globally optimal solution is not zero. In practice, however, this theoretical result is not tremendously useful.

8.1.4 *Fast Ant System*

One of the disadvantages of numerous frameworks based on artificial ants is their large number of parameters and the difficulty of tuning them. This is the reason why we have not presented *Ant systems* (AS [1]) or *Ant Colony System* (ACO [3]) in detail. In addition, it can be challenging to design an incremental cost function providing pertinent results. An example is the quadratic assignment problem. Since any pair of elements contributes to the fitness function, the ultimate element to include can contribute significantly to the quality of the solution. Conversely, the first item placed does not incur any cost. This is why a simplified framework called *FANT* (for Fast Ant System) has been proposed.

In addition to the number of iterations, I_{max}, the user of this framework must only specify another parameter, τ_b. It corresponds to the reinforcement of the artificial pheromone trails. This reinforcement is systematically applied to the elements of the best solution found so far at each iteration. The reinforcement of the traces associated with the elements of the solution constructed at the current iteration, τ_c, is a self-adaptive parameter. Initially, this parameter is set to 1. When over-learning is detected (the best solution is again generated), τ_c is incremented, and all trails are reset to τ_c. This implements the oblivion process and increases the diversity of the solutions generated.

If the best solution has been improved, then τ_c is reset to 1 to give more weight to the elements constituting this improved solution. Ultimately, FANT incorporates a local search method. As mentioned above, it has indeed been noticed that the sole construction mechanism often produces bad quality solutions. Algorithm 8.2 provides the FANT framework.

Figure 8.2 illustrates the FANT behavior on a TSP instance with 225 cities. In this experiment, the value of τ_b was fixed to 50. This figure provides the number of edges different from the best solution found so far, before and after calling the improvement procedure.

A natural implementation of the trails for the TSP is to use a matrix τ rather than a vector. Indeed, an element e of a solution is an edge $[i, j]$, defined by its two incidents vertices. Therefore the value τ_{ij} is the a posteriori interest to have the edge $[i, j]$ in a solution. The initialization of this trail matrix and its update may therefore be implemented with the procedures described by Code 8.2.

Algorithm 8.2: FANT framework. Most of the lines of code are about automatically adjusting the weight τ_c assigned to the newly built solution against the τ_b weight of the best solution achieved so far. If the latter is improved or if over-learning is detected, the trails are reset

Input: Set E of elements constituting a solution; fitness function f to minimize, parameters I_{max}, τ_b and improvement method $a(\cdot)$
Result: Solution s^*

$f^* \leftarrow$
$\tau_c \leftarrow 1$
for $\forall e \in E$ **do**
    $\tau_e \leftarrow \tau_c$
for I_{max} *iterations* **do**
    Initialize s to a partial, trivial solution
    $R \leftarrow E$ // Elements that can be added to s
    while $R \neq \varnothing$ **do**
        Randomly choose $e \in R$ with a probability proportionnal to τ_e
        $s \leftarrow s \cup e$
        From R, remove the elements that cannot be added any more to s
    $s' \leftarrow a(s)$ // Find the local optimum s' associated with s
    if $s' = s^*$ **then** manage over-learning
        $\tau_c \leftarrow \tau_c + 1$ // More weight to the newly constructed solutions
        for $\forall e \in E$ **do** Erase all trails
            $\tau_e \leftarrow \tau_c$
    if $f^* > f(s_k)$ **then** manage best solution improvement
        $f^* \leftarrow f(s_k)$
        $s^* \leftarrow s_k$ // Update best solution
        $\tau_c \leftarrow 1$ // Give minimum weight to the newly constructed solutions
        for $\forall e \in E$ **do** Erase all trails
            $\tau_e \leftarrow \tau_c$
    for $\forall e \in s'$ **do** reinforce the trails associated with the current solution
        $\tau_e \leftarrow \tau_e + \tau_c$
    for $\forall e \in s^*$ **do** reinforce the trails associated with the best solution
        $\tau_e \leftarrow \tau_e + \tau_b$

The core of an ant heuristic is the construction of a new solution exploiting artificial pheromones. Code 8.1 provides a procedure not exploiting the a priori interest (an incremental cost function) of the elements constituting a solution. In this implementation, the departure city is the first of a random permutation p. At iteration i, the i first cities are definitively chosen. At that time, the next city is selected with a probability proportional to the trail values of the remaining elements.

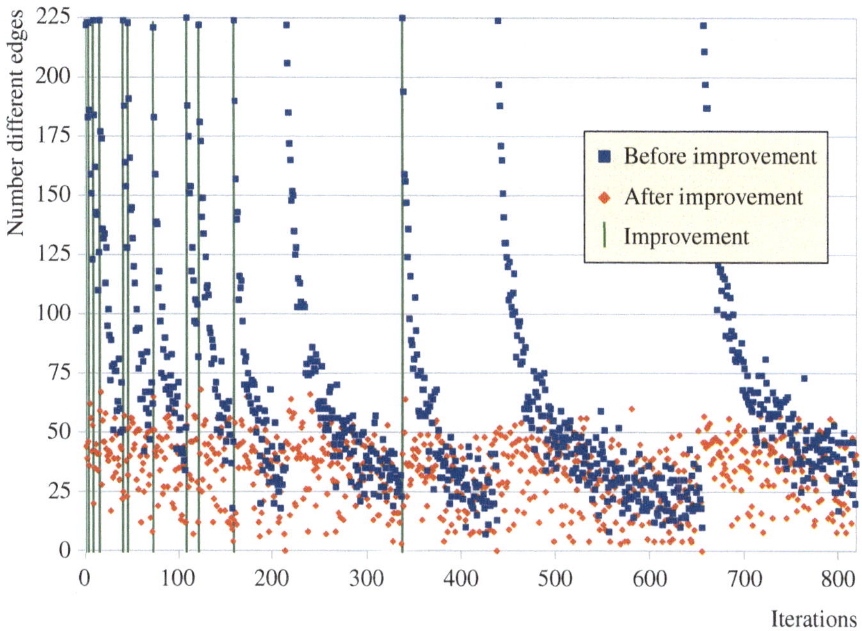

Fig. 8.2 FANT behaviour on a TSP instance with 225 cities. For each iteration, the diagram provides the number of edges different from the best solution found by the algorithm, before and after calling the ejection chain local search. Vertical lines indicate improvements in the best solution found. In this experiment, the last of these improvements corresponds to the optimal solution

Code 8.1 generate_solution_trail.py Implantation of the generation of a permutation only exploiting the information contained in the pheromone trails

```
from random_generators import unif                               # Listing 12.1
from tsp_utilities import tsp_length                             # Listing 12.2

######### Building a solution using artificial pheromone trails
def generate_solution_trail(d,                                 # Distance matrix
                            tour,                     # Tour produced by the ant
                            trail):                         # Pheromone trails
    n = len(tour)
    for i in range(1, n - 1):
        total = 0
        for j in range(i + 1, n):
            total += trail[tour[i - 1]][tour[j]]
        target = unif(0, total - 1)
        j = i
        total = trail[tour[i - 1]][tour[j]]
        while total < target:
            total += trail[tour[i - 1]][tour[j + 1]]
            j += 1
        tour[j], tour[i] = tour[i], tour[j]
    return tour, tsp_length(d, tour)
```

Once the three procedures given by the Codes 8.1 and 8.2 as well as an improvement procedure are available, the implementation of FANT is very simple. Such an implantation, using an ejection chain local search, is given by Code 8.3

Code 8.2 init_update_trail.py Implementation of the trail matrix initialization and update for the FANT method applied to a permutation problem. If the solution just generated is the best previously found, trails are reset. Otherwise, the trails are reinforced both with the current solution and the best one

```
######## (Re-)initialize all trails
def init_trail(initial_value,                    # Initial value for all trails
               trail):                                        # Pheromone trails

    n = len(trail[0])
    for i in range(n):
        for j in range(n):
            trail[i][j] = initial_value
    for i in range(n):
        trail[i][i] = 0
    return trail

######### Updating trail values
def update_trail(tour,                          # Last solution generated by an ant
                 global_best,                                # Global best solution
                 exploration,                            # Reinforcement of last solution
                 exploitation,                    # Reinforcement of global best solution
                 trail):                                        # Pheromone trails

    if tour == global_best:
        exploration += 1                          # Give more weight to exploration
        trail = init_trail(exploration, trail)
    else:
        for i in tour:
            n = len(trail[0])
            trail[tour[i]][tour[(i + 1) % n]] += exploration
            trail[global_best[i]][global_best[(i + 1) % n]] += exploitation
    return trail, exploration
```

Code 8.3 tsp_FANT.py FANT for the TSP. The improvement procedure is given by Code 12.3

```
from random_generators import rand_permutation                # Listing 12.1
from generate_solution_trail import *                         # Listing  8.1
from init_update_trail import *                               # Listing  8.2
from tsp_LK import tsp_LK                                     # Listing 12.3

######### Fast Ant System for the TSP
def tsp_FANT(d,                                               # Distance matrix
             exploitation,          # FANT Parameters: global reinforcement
             iterations):                # number of solution to generate

    n = len(d[0])
    best_cost = float('inf')
    exploration = 1
    trail = [[-1] * n for _ in range(n)]
    trail = init_trail(exploration, trail)
    tour = rand_permutation(n)
    for i in range(iterations):
        # build solution
        tour, cost = generate_solution_trail(d, tour, trail)
        # improve built solution witho a local search
        tour, cost = tsp_LK(d, tour, cost)
        if cost < best_cost:
            best_cost = cost
            print('FANT {:d} {:d}'.format(i+1, cost))
            best_sol = list(tour)
            exploration = 1              # Reset exploration to lowest value
            trail = init_trail(exploration, trail)
        else:
            # pheromone trace reinforcement - increase memory
            trail, exploration = update_trail(tour, best_sol,
                                      exploration, exploitation, trail)
    return best_sol, best_cost
```

8.2 Vocabulary Building

Vocabulary building is a more global learning method than artificial ant colonies. The idea is to memorize fragments of solutions, which are called words, and to construct new solutions from these fragments. Put differently, one has a dictionary used to build a sentence attempt in a randomized way. A repair/improvement procedure makes this solution attempt feasible and increases its quality. Finally, this new solution sentence is fragmented into new words that enrich the dictionary.

This method has been proposed in [4] and is not yet widely used in practice, although it has proved efficient for a number of problems. For instance, the method can be naturally adapted to the vehicle routing problem. Indeed, it is relatively easy to construct solutions with tours similar to those of the most efficient solution known. This is illustrated in Fig. 8.3.

By building numerous solutions using randomized methods, the first dictionary of solution fragments can be acquired. This is illustrated in Fig. 8.4.

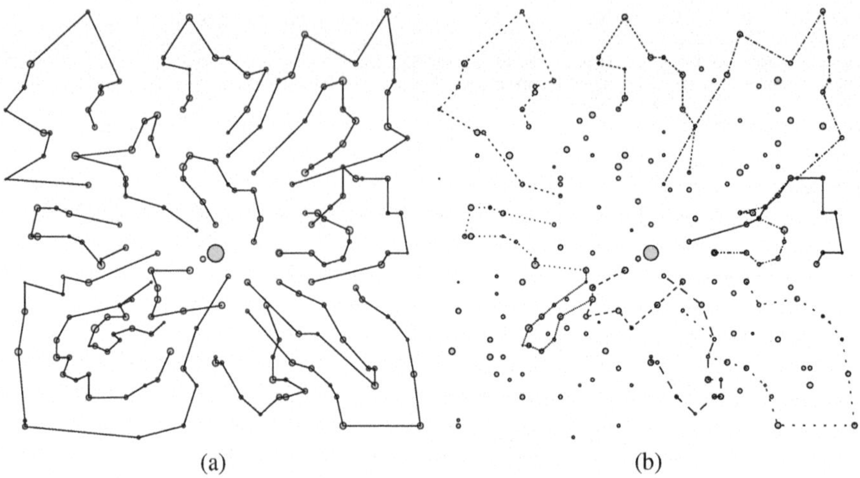

Fig. 8.3 (a) The optimal solution to a VRP instance. (b) A few tours quickly obtained with a taboo search. We notice great similarities between the latter and those of the optimal solution

Fig. 8.4 Fragments of solutions (vehicle routing tours) constituting the dictionary. A partial solution is built by randomly selecting a few of these fragments (indicated in color)

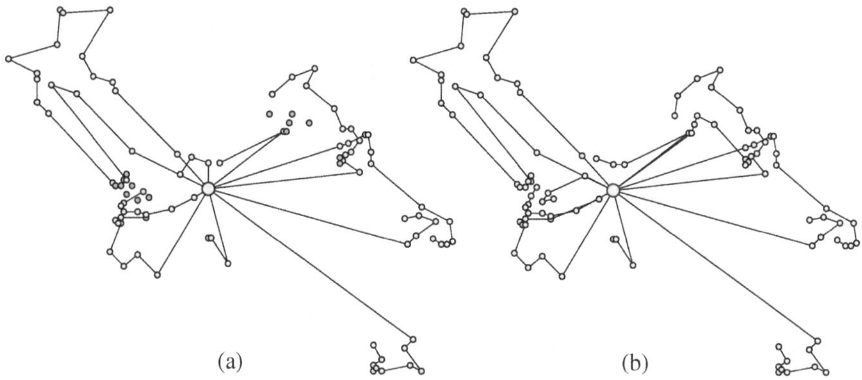

Fig. 8.5 (**a**) A sentence attempt is constructed by randomly selecting a few words from dictionary (**b**). This attempt is completed and improved

Once an initial dictionary has been constructed, solution attempts are built, for instance, by selecting a subset of tours that do not contain common customers. This solution is not necessarily feasible. Indeed, during the construction process, the dictionary might not include tours only containing customers not yet covered. Therefore, it is necessary to repair this solution attempt, for instance, by means of a method similar to that used to produce the first dictionary but starting with the solution attempt. This phase of the method is illustrated in Fig. 8.5. The improved solution is likely to contain tours that are not yet in the dictionary. These are included to enrich it for subsequent iterations.

The technique can be adapted to other problems, like the TSP. In this case, the dictionary words can be edges appearing in a tour. Figure 8.6 shows all the edges present in more than two-thirds of 100 tours obtained by applying a local search starting with a random solution. The optimal solution to this problem is known. Hence, it is possible to highlight the few edges frequently obtained that are not part of the optimal solution. Interestingly, nearly 80% of the edges of the optimal solution have been identified by initializing the dictionary with a basic improvement method.

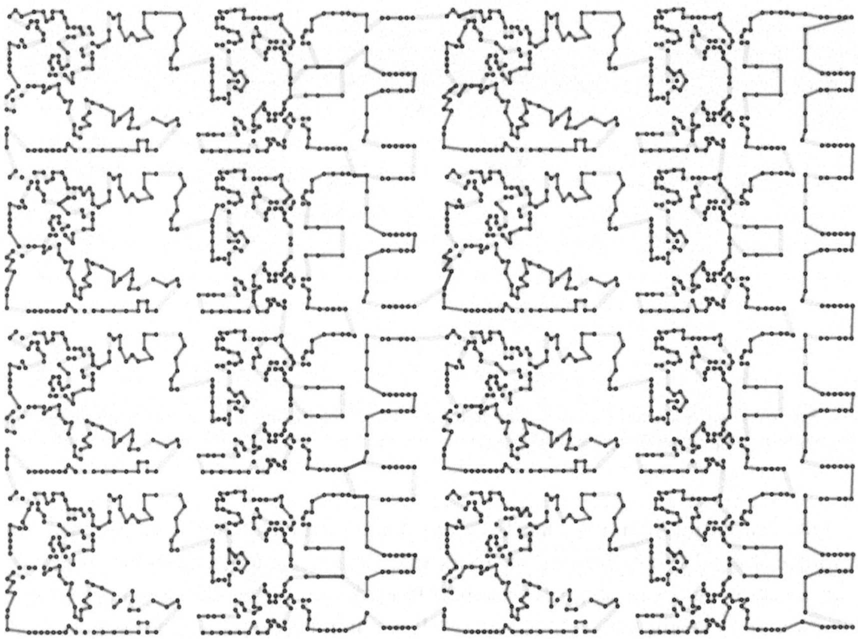

Fig. 8.6 An optimal solution (light color) and fragments of tours constituting an initial dictionary for the TSP instance pr2392. The fragments are obtained by repeating 100 local searches starting with random solutions and only retaining the edges appearing in more than 2/3 of the local optima. Interestingly, almost all these edges belong to an optimal solution. The few edges that are not part of it are highlighted (darkest color)

Problems

8.1 Artificial Ants for Steiner Tree

For the Steiner tree problem, how to define the trails of an artificial ant colony? Describe how these trails are exploited.

8.2 Tuning the FANT Parameter

Determine good values for the parameter τ_b of the `tsp_FANT` method provided by Code 8.3 when the latter performs 300 iterations. Consider the *TSPLIB* instance tsp225.

8.3 Vocabulary Building for Graph Coloring

Describe how vocabulary construction can be adapted to the problem of coloring the vertices of a graph.

References

1. Colorni, A., Dorigo, M., Maniezzo, V.: Distributed optimization by ant colonies. In: Actes de la première conférence européenne sur la vie artificielle, pp. 134–142. Elsevier, Paris (1991)
2. Deneubourg, J., Goss, S., Pasteels, J., Fresneau, D., Lachaud, J.: Self-organization mechanisms in ant societies (II): Learning in foraging and division of labor. In: Pasteels, J., et al. (eds.) From Individual to Collective Behavior in Social Insects, Experientia supplementum, vol. 54, pp. 177–196. Birkhäuser, Basel (1987)
3. Dorigo M., Gambardella L.M.: Ant Colony System : A Cooperative Learning Approach to the Traveling Salesman Problem. IEEE Trans. Evol. Comput. **1**(1), 53–66 (1997). https://doi.org/10.1109/4235.585892
4. Glover, F.: Tabu search and adaptive memory programming — advances, applications and challenges. In: Barr, R.S., Helgason, R.V., Kennington, J.L. (eds.) Interfaces in Computer Science and Operations Research: Advances in Metaheuristics, Optimization, and Stochastic Modeling Technologies, pp. 1–75. Springer, Boston (1997). https://doi.org/10.1007/978-1-4615-4102-8_1
5. Stützle, T., Hoos, H.H.: MAX MIN Ant system. Future Gener. Comput. Syst. **16**(8), 889–914 (2000). https://doi.org/10.1016/S0167-739X(00)00043-1

Chapter 9
Local Search Learning

Local searches play an essential role in metaheuristics. Virtually, all efficient heuristic methods incorporate a local search. Moreover, metaheuristics are sometimes defined as a master process guiding a local search. In Chapter 5, we have already seen some basic neighborhood adaptation techniques, in particular its limitation by the list of candidate moves, granular search and its extension by filter-and-fan search, and ejection chains.

Most randomized methods reviewed in Chapter 7 are dedicated to local search extensions. They are not implementing a learning process. They only memorize the best solution found so far or statistics for self-calibrating the parameters. This allows taking the unit of measurement of the fitness function into account. The following step in the sophistication of metaheuristics is to learn to locally modify solutions to a problem. Among the popular techniques, taboo search (also written tabu search) offers many strategies and various local search learning mechanisms. This chapter reviews the basic mechanisms. Other strategies are proposed in the book of Glover and Laguna [5] and take a more natural place in other chapters of the present book.

9.1 Taboo Search

Proposed by Fred Glover in 1986, the key idea of taboo search is to explore the solution space with a local search beyond local optima [2–4]. This implies designing a mechanism to prevent the cycling phenomenon, the fact of entering a cycle where a limited subset of solutions is repeatedly visited. The simplest concept to imagine is to memorize all the solutions which have been successively encountered during a local search but preventing the latter from choosing a neighbor solution that has already been visited. The visited solutions thus become taboo.

This concept is simple but cumbersome to implement: imagine that a local search can require millions of iterations, which means memorizing the same number of

É. D. Taillard, *Design of Heuristic Algorithms for Hard Optimization*, Graduate Texts in Operations Research, https://doi.org/10.1007/978-3-031-13714-3_9

solutions. Each neighbor solution must be checked to ensure that it has not already been explored. Knowing that a neighborhood can contain thousands of solutions, we quickly realize the hopelessness of this way of doing things, either because of the memory space needed to store the visited solutions or because of the computational effort to compare neighbor solutions.

9.1.1 *Hash Table Memory*

A simple technique to implement an approximation of this principle of prohibiting previously visited solutions is to use a hash table. An integer value $h(s)$ is associated with each solution s of the problem. If we visit the solution s_i at iteration i of the search, we store the value i in the entry $h(s_i) \mod m$ of an array T of m integers. Thus, the value t_k of the kth entry of the table T indicates at which iteration a solution whose hash value k (modulo m) has been visited.

The h function is generally not bijective over the set of solutions to the problem, so various solutions can have the same hash value. Indeed, the size m of the array T must be limited due to the available memory. This technique is, therefore, an approximation of the concept of prohibiting solutions already visited. Indeed, not only the latter is prohibited but also all those that have the same hash value. Moreover, since the value of m is limited, we cannot forever forbid returning to a solution of a given hash value. After m iterations at the latest, all the solutions would be prohibited.

It is therefore necessary to implement a key feature of the learning process: oblivion. These considerations lead us to introduce the key parameter of a taboo search: the taboo duration, sometimes referred to as the *taboo list length.*

9.1.1.1 Hash Functions

The choice of a hash function to implement a taboo search is not very difficult. In some cases, the value of the fitness function is perfect, especially when the neighborhood includes many moves at zero cost (plateaus). Indeed, taboo search chooses the best move allowed at each iteration. Hence, neutral changes make learning difficult. Being on a plateau, the choice of one or the other neighbor is problematic and cycling can occur. In case the fitness function admits an extensive range of values, prohibiting during a number of iterations to return to a given fitness value allows, in many cases, to break the local optimum structure and discover another one.

A general hash function is as follows. Let us use the notation introduced in Chapter 4 devoted to constructive methods. A solution is composed of elements $e \in E$. Each of them is associated with an integer value z_e. These values are randomly generated at the beginning of the algorithm. The hash value of a solution s is provided by $h(s) = \sum_{e \in s} z_e$. A more sophisticated hash technique using multiple tables is discussed in [6]. It makes it possible to obtain the equivalent of a very large table, while limiting the memory space.

9.1.2 *Taboo Moves*

Prohibition based on a hash function is uncommon in taboo search implementations. Frequently, one prohibits some moves or solutions with certain features. To be concrete, consider the example of the symmetric TSP.

A 2-opt move can be characterized by the pair $[i, j]$ which consists in replacing the edges $[i, s_i]$ and $[j, s_j]$ of the current solution s by the edges $[i, j]$ and $[s_i, s_j]$. One assumes here that the solution is provided by the "successor" of each city and that the city j comes "after" the city i when traveling in the order given by s. If the move $[i, j]$ is carried out at an iteration, one can prohibit the reverse move $[i, s_i]$ during the following iterations. This is a direct prohibition based on the opposite of a move.

After performing the move $[i, j]$, another possibility is to indirectly prohibit the moves leading to a solution containing both edges $[i, s_i]$ and $[j, s_j]$.

By abuse of language, let m^{-1} denote the inverse of a move, or a feature of a solution that is forbidden after performing the move m of a neighborhood characterized by a set M of moves. Although $(s \oplus m) \oplus m^{-1} = s$, there may be various ways to define m^{-1}. Since the size of the neighborhood is limited, it is necessary to relax the taboo status of a move after relatively few iterations. Therefore, the taboo list is frequently presented as a *short-term memory*. The most basic taboo search framework is given by Algorithm 9.1.

9.1.2.1 Implementation of Taboo Status

If the neighborhood size is not too large, it can be stored, for each move, the iteration from which it can be used again. Let us immediately illustrate such an implementation for the following knapsack instance with nine variables.

$$\begin{aligned} \max r &= 12s_1 + 10s_2 + 9s_3 + 7s_4 + 4s_5 + 8s_6 + 11s_7 + 6s_8 + 13s_9 \\ \text{Subject } & 10s_1 + 12s_2 + 8s_3 + 7s_4 + 5s_5 + 13s_6 + 9s_7 + 6s_8 + 14s_9 \leqslant 45 \\ \text{to: } & s_i \in \{0, 1\} \quad (i = 1, \dots, 9) \end{aligned} \tag{9.1}$$

A solution s of this problem is a 0—1 vector, with $s_i = 1$ if the object i is chosen and $s_i = 0$ otherwise. Each object occupies a certain volume in the knapsack and the latter possesses a global volume of 45. An elementary neighborhood for this problem is to alter the value of a unique variable of s.

The taboo conditions can be stored as a vector t of integers with t_i giving the iteration number at which the variable s_i can revert to a previous value. Initially,

Algorithm 9.1: Elementary taboo search framework

Input: Solution s, set M of moves, fitness function $f(\cdot)$ to minimize, parameters I_{max}, d.
Result: Improved solution s^*

$s^* \leftarrow s$
for I_{max} *iterations* **do**
  $best\_neighbor\_value \leftarrow \infty$
  forall $m \in M$ *(such that m (or* $s \oplus m$*) is not marked as taboo)* **do**
    if $f(s \oplus m) < best\_neighbor\_value$ **then**
      $best\_neighbor\_value \leftarrow f(s \oplus m)$
      $m^* \leftarrow m$
  if $best\_neighbor\_value < \infty$ **then**
    Mark $(m^*)^{-1}$ (or s) as taboo for the next d iterations
    $s \leftarrow s \oplus m^*$
    if $f(s) < f(s^*)$ **then**
      $s^* \leftarrow s$
  else
    Error message: d too large: no move allowed!

$t = 0$: at the first iteration, all variables can be modified. For this small instance, let us assume a taboo duration of $d = 3$. The initial solution can be set to $s = 0$, which represents the worst feasible solution to the problem. Table 9.1 gives the evolution of a taboo search for this small instance.

Unsurprisingly, object 9 is put in the knapsack at the first iteration. Indeed, this object has the largest value. At the end of iteration 1, it is forbidden to set $s_9 = 0$ again up to the iteration $t_9 = 4 = 1 + 3$. As long as there is room in the knapsack, taboo search behaves like a greedy constructive algorithm. At iteration 4, it reaches the first local optimum $s = (1, 1, 0, 0, 0, 0, 1, 0, 1)$ of value $r = 46$.

Table 9.1 Evolution of an elementary taboo search for ten iterations for the knapsack instance 9.1. This search forbids changing again a given variable for $d = 3$ iterations

Iteration number	Variable modified	Modified solution	Fitness value	Volume used	Taboo status
1	s_9	(0, 0, 0, 0, 0, 0, 0, 0, 1)	13	14	(0, 0, 0, 0, 0, 0, 0, 0, 4)
2	s_7	(1, 0, 0, 0, 0, 0, 0, 0, 1)	25	24	(5, 0, 0, 0, 0, 0, 0, 0, 4)
3	s_7	(1, 0, 0, 0, 0, 0, 1, 0, 1)	36	33	(5, 0, 0, 0, 0, 0, 6, 0, 4)
4	s_2	(1, 1, 0, 0, 0, 0, 1, 0, 1)	46	45	(5, 7, 0, 0, 0, 0, 6, 0, 4)
5	s_9	(1, 1, 0, 0, 0, 0, 1, 0, 0)	33	31	(5, 7, 0, 0, 0, 0, 6, 0, 8)
6	s_3	(1, 1, 1, 0, 0, 0, 1, 0, 0)	42	39	(5, 7, 9, 0, 0, 0, 6, 0, 8)
7	s_8	(1, 1, 1, 0, 0, 0, 1, 1, 0)	48	45	(5, 7, 9, 0, 0, 0, 6, 10, 8)
8	s_2	(1, 0, 1, 0, 0, 0, 1, 1, 0)	38	33	(5, 11, 9, 0, 0, 0, 6, 10, 8)
9	s_4	(1, 0, 1, 1, 0, 0, 1, 1, 0)	45	40	(5, 11, 9, 12, 0, 0, 6, 10, 8)
10	s_5	(1, 0, 1, 1, 1, 0, 1, 1, 0)	49	45	(5, 11, 9, 12, 13, 0, 6, 10, 8)

At iteration 5, an object is removed, because the knapsack is dead full. Only object 9 can be removed due to taboo conditions. As a result, the fitness function decreases from $r = 46$ to $r = 33$, but space is freed up in the knapsack. At iteration 6, the best move would be to add object 9, but this move is taboo. It would correspond to return to the solution visited at iteration 4.

The best authorized move is therefore to add object 3. Then, at the subsequent iteration the object 8 is added, leading to a new local optimum $s = (1, 1, 1, 0, 0, 0, 1, 1, 0)$ of value $r = 48$. The knapsack is again completely full. At iteration 8, it is necessary to remove an object, setting $s_2 = 0$. The place thus released makes it possible to add the objects 4 and 5, discovering a solution $s = (1, 0, 1, 1, 1, 0, 1, 1, 0)$ even better than both local optima previously found.

For the TSP, the type of restrictions described above may be implemented using a matrix T whose entry t_{ij} provides the iteration from which we can again perform a move where the edge $[i, j]$ belongs to the tour. This principle extends to any combinatorial problem for which we search for an optimal permutation.

9.1.2.2 Taboo Duration

In the previous example, the taboo duration was set to three iterations. This value may seem arbitrary. If the taboo conditions are removed (duration set to zero), the search enters a cycle. Once a local optimum is reached, an object is removed and added again in the next iteration. The maximum taboo duration is clearly limited by the neighborhood size: indeed, the search performs all the moves of the neighborhood and then remains blocked. At that time, they are all prohibited.

These two extreme cases lead to inefficient searches—a zero taboo duration is equivalent to learning nothing; a very high duration implies poor learning. Consequently, we have to achieve a sensible compromise for the taboo duration. Therefore, this duration must be learned for the problem instance treated. Figure 9.1 illustrates this phenomenon for Euclidean TSP instances of size $n = 100$ randomly, uniformly distributed in a square. The taboo search performs $I_{max} = 1000$ iterations.

Battiti and Tecchiolli [1] proposed a learning mechanism called *reactive taboo search*. All the solutions visited by the search are memorized. They can be stored in an approximate way, employing the hash technique presented in Section 9.1.1.1. The search starts with a restricted taboo duration. If the search visits a solution again, then the duration is increased. If the search does not revisit any of the solutions during a relatively significant number of iterations, then the taboo duration is diminished.

This last condition seems strange. Why should we force the search to return to previously explored solutions? The explanation is as follows: if the taboo duration is sufficient to avoid the cycling phenomenon, it also means we are forbidden to visit some good solutions because of the taboo status. We are therefore likely to ignore high-quality solutions.

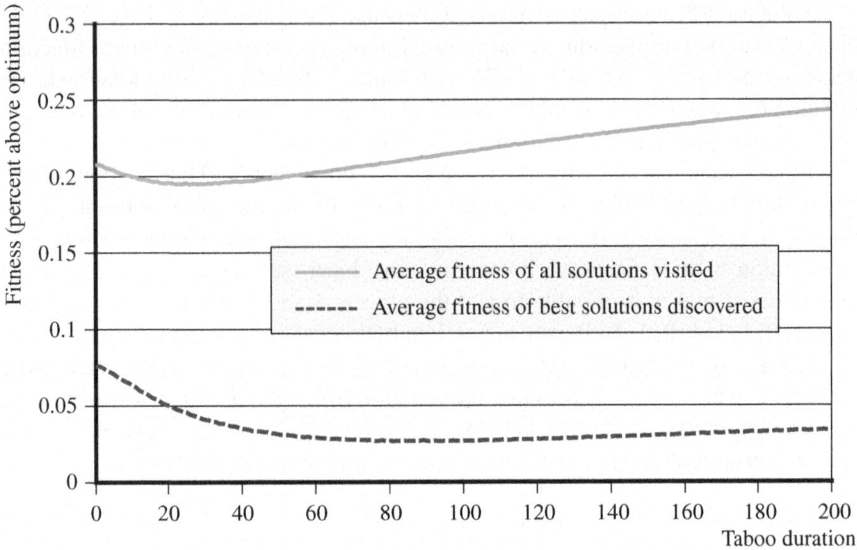

Fig. 9.1 Influence of the taboo duration for Euclidean TSP with 100 cities. A short duration allows visiting better quality solutions, on average. But the search cannot escape from local optima. Hence, the quality of the best solutions found is not excellent. Conversely, if the taboo duration is too high, the average solution quality decreases, as well as that of the best solutions discovered. In this case, a reasonable compromise seems to be a taboo duration around the instance size. More generally, the square root of the neighborhood size seems appropriate

It is therefore necessary to find a taboo duration long enough to avoid cycling but as short as possible so as not to prohibit good moves. This is precisely the purpose of reactive taboo search.

However, this learning technique only repels the problem. Indeed, the user must determine another parameter which is the number of iterations without revisiting a solution, triggering the taboo duration decrease. In addition, it requires the implementation of a storage mechanism for all visited solutions, which can be cumbersome to implement.

Another technique for choosing low taboo durations while strongly preventing the cycling phenomenon is to randomly set it at each iteration. A classic method is to select the taboo duration at random between a minimum duration d_{min} and a maximum value $d_{max} = d_{min} + \Delta$.

To create Fig. 9.2, 500 QAP instances of size $n = 12$ with known optimal solution have been generated. For each instance, we have performed a taboo search with a considerable number of iterations (for instances that small) with all possible parameters (d_{min}, Δ). The number of optimal solutions found for each couple was then counted. If the search succeeds in finding all the 500 optimal solutions, the average number of iterations needed to reach the optimum is recorded.

With a deterministic taboo duration ($\Delta = 0$), it was never possible to achieve all the optimal solutions, even with a relatively large duration. Conversely, with

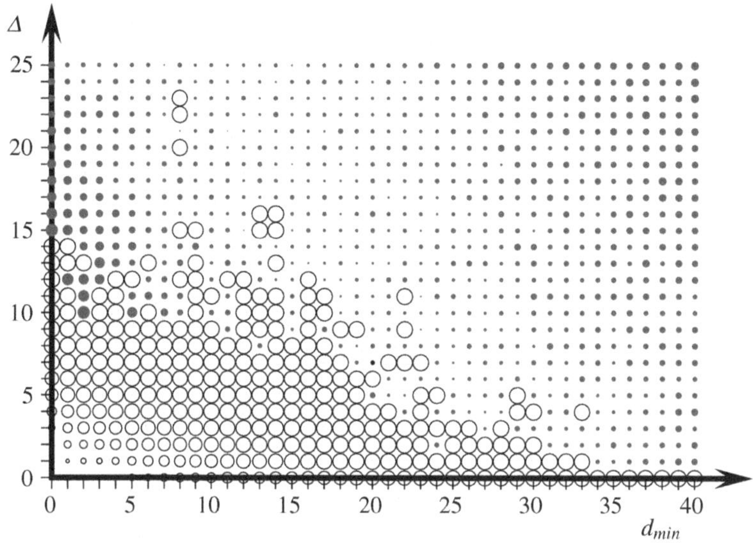

Fig. 9.2 Taboo duration randomly generated between d_{min} and $d_{min} + \Delta$. An empty circle indicates that taboo search has been unable to systematically find the optimum of 500 QAP instances. The circle size is proportional to the number of optimum found (the larger, the better). A filled disc indicates that the optimum has been systematically found. The disk size is proportional to the average number of iterations required for obtaining the optimum (the smaller, the better)

low minimum durations and a random variation equals to the size of the problem, the optimum is systematically obtained. Moreover, the optimum is reached with relatively few iterations.

Figure 9.3 reproduces a similar experiment for the TSP. It provides the solution quality obtained for a small instance for any couple (d_{min}, Δ). We observe similarities with Fig. 9.2. A random taboo duration proportional to half the instance size is a reasonable compromise.

9.1.2.3 Aspiration Criterion

The unconditional prohibition of a move can cause unwanted situations. For instance, one can skip an improvement of the best solution found. Thus, Line 4 of Algorithm 9.1 is modified, and if the move m allows achieving a solution better than s^*, it is retained. In the literature, this is referred to as an *aspiration criterion*. Other less trivial aspiration criteria can be imagined, in particular to implement a long-term memory.

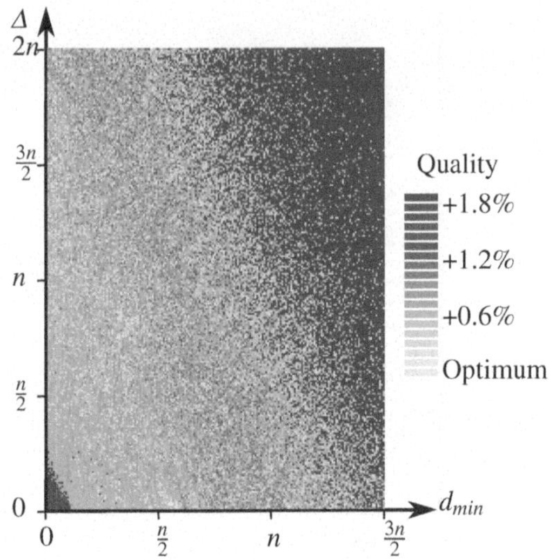

Fig. 9.3 Quality of the solutions obtained with a taboo search where the taboo duration is randomly chosen between d_{min} and $d_{min} + \Delta$ for a classical instance with $n = 127$ cities. The method performs $10n$ iterations starting from a deterministic greedy nearest neighbor tour. For all values of d_{min} between 0 and $1.5n$ and Δ between 0 and $2n$, we launched a search and represented the solution quality by a color (% above optimum)

9.2 Strategic Oscillations

Forbidding the inverse of moves recently performed implements a short-term memory. This mechanism can be very efficient for instances of moderate size. By cons, if we address more complex problems, this sole mechanism is not sufficient. A search strategy that has been proposed in the context of taboo search is to alternate *intensification and diversification* phases.

The goal of intensification is to thoroughly examine a limited portion of the search space, maintaining solutions that possess a globally similar structure. Once all the attractive solutions of this portion are supposedly discovered, the search has to go elsewhere. Put differently, the search is diversified by altering the structure of the solution. The search intensification can be implemented with a short duration taboo list.

9.2.1 Long-Term Memory

Implementing a diversification mechanism supposes to include a long-term memory. Several techniques have been proposed to achieve that.

9.2.1.1 Forced Moves

The most certain and convenient way to break the structure of a solution is to perform moves that have never been selected during many iterations. With a basic taboo search memorizing the iteration from which each move can again be performed, the implementation of this form of long-term memory is virtually free. Indeed, if the iteration number stored for a move is considerably smaller than the current iteration, then this move has not been selected for a long time.

It is thus possible to force the use of this modification, regardless of the quality of the solution to which it leads. This mechanism requires a new parameter, K, representing the number of iterations from which a never chosen move is forced. Naturally, this parameter must be larger than the size of the neighborhood; otherwise the search degenerates, performing only forced moves. If several moves are to be forced at a given iteration, one is chosen arbitrarily. The others will be forced in subsequent iterations. This type of long-term memory represents a kind of aspiration criterion, introduced in the previous section.

9.2.1.2 Penalized Moves

A weakness of taboo search with very short-term memory is that it only makes small changes. To illustrate this on the TSP, such a search will "knit" a small knot on a tour that was locally optimal, then another elsewhere and so on until the taboo condition drops. At that point, the search unknits the first knot. This situation is illustrated in Fig. 9.4.

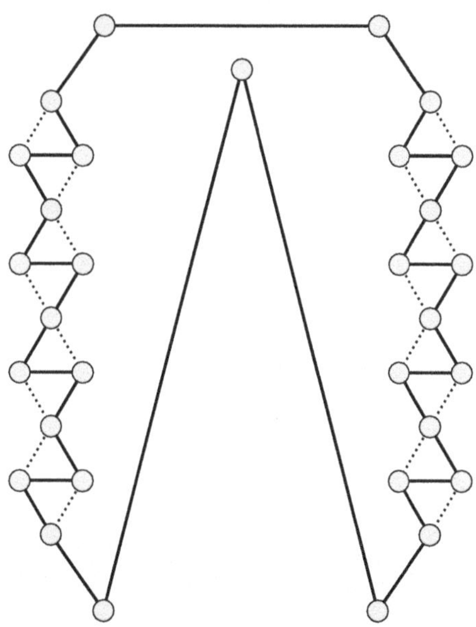

Fig. 9.4 A basic taboo search with a short-term memory can enter cycling with this 2-optimal tour. Indeed, this solution belongs to a plateau. There are eight moves not changing the tour length (dotted lines). With a taboo duration shorter than eight, the search repeatedly chooses one of these moves

To avoid this behavior, an idea is to store the number of times f_m a move m was chosen and limit its use. During the move evaluation, a penalty $F \cdot f_m$ is added. The proportionality factor F is a new parameter of the technique that must be tuned. Naturally, an aspiration criterion must be used in conjunction with this mechanism. Indeed, the search should nevertheless be allowed choosing a heavily penalized move leading to an improvement of the best solution known.

Code 9.1 provides a taboo search implementation for the TSP, based on the 2-opt neighborhood. Two types of memories are employed: a short-term conventional memory that prevents moves from reintroducing both edges that have been recently removed and a long-term memory that counts the number of times each edge has been inserted in the solution.

A move is penalized proportionally to the number of times the concerned edges have been introduced in the solution. A move is forbidden if both edges have recently been removed from the solution (eventually at different iterations). Ultimately, a move is aspired if it improves on the best solution achieved so far.

Code 9.1 tsp_TS.py Taboo search implementation for the TSP

```
from random_generators import *                                    # Listing 12.1

######### Taboo Search for the TSP, based on 2-opt moves
def tsp_TS(d,                                                    # Distance matrix
           tour,                                           # Intital tour provided
           length,                                         # Length of initial tour
           iterations,                            # Number of tabu search iterations
           min_tabu_duration,                              # Minimal tabu duration
           max_tabu_duration,
           F):                   # Factor for penalizing moves repeatedly performed

    n = len(tour)
    tabu = [[0] * n for _ in range(n)]                                   # Tabu list
    count = [[0] * n for _ in range(n)]                                 # Move count
    best_tour = tour[:]
    best_length = length
    for iteration in range(0, iterations):
        delta_penalty = float('inf')
        ir = jr = -1                          # Cities retained for performing a move

        # Find best move allowed or aspired
        for i in range(n - 2):
            j = i + 2
            while j < n and (i > 0 or j < n - 1):
                delta =  d[tour[i]][tour[j]]   + d[tour[i+1]][tour[(j+1) % n]]\
                    -d[tour[i]][tour[i + 1]]   - d[tour[j]][tour[(j + 1) % n]]
                penality =  F * (count[tour[i]][tour[j]]
                                 + count[tour[i + 1]][tour[(j + 1) % n]])
                # Conditions for accepting a candidate move
                better = delta + penality < delta_penalty
                allowed = tabu[tour[i]][tour[j]] <= iteration \
                          or  tabu[tour[i + 1]][tour[(j + 1) % n]] <= iteration
                aspirated = length + delta < best_length

                if better and (allowed or aspirated):
                    delta_penalty = delta + penality
                    ir, jr = i, j
                j += 1                                               # Next neighbor

        # Perform retained move
        if delta_penalty < float('inf'):
            tabu[tour[ir]][tour[ir + 1]] = tabu[tour[jr]][tour[(jr + 1) % n]] \
            = tabu[tour[ir + 1]][tour[ir]] = tabu[tour[(jr+1) % n]][tour[jr]] \
            = unif(min_tabu_duration, max_tabu_duration) + iteration

            count[tour[ir]][tour[ir + 1]] += 1
            count[tour[jr]][tour[(jr + 1) % n]] += 1
            count[tour[ir + 1]][tour[ir]] += 1
            count[tour[(jr + 1) % n]][tour[jr]] += 1

            length += d[tour[ir]][tour[jr]] + d[tour[ir+1]][tour[(jr+1) % n]]\
                     -d[tour[ir]][tour[ir+1]] - d[tour[jr]][tour[(jr+1) % n]]

            for k in range((jr - ir) // 2):
                tour[k + ir + 1], tour[jr - k] = tour[jr - k], tour[k + ir + 1]
        else:
            print('All moves are forbidden tabu list too long')
        if best_length > length:                          # is there an improvement?
            best_length = length
            best_tour = tour[:]
            print('TS {:d} {:d}'.format(iteration+1, best_length))
    return best_tour, best_length
```

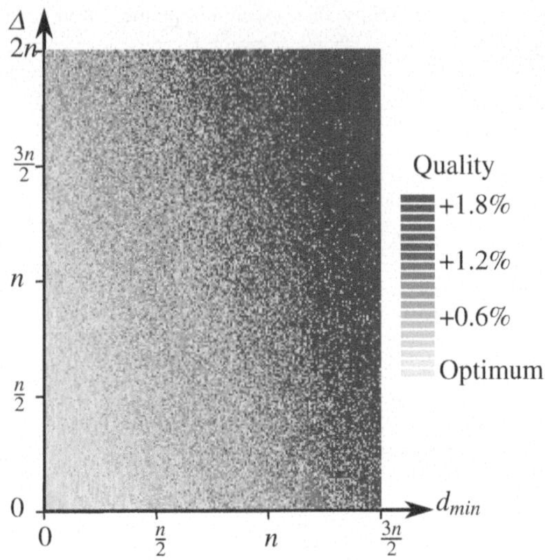

Fig. 9.5 Same diagram as Fig. 9.3 but with a taboo search managing a long-term memory. Frequently performed moves are penalized. The value of the penalty is $F \cdot n_e$, where n_e is the number of times the edge e has been included in or removed from the tour and the value of F is the average length of an edge divided by the instance size

Figure 9.5 illustrates the quality of this taboo search performing $10n$ iterations on a TSP instance with $n = 127$ cities. The search implements the penalty mechanism based on the frequency of moves. Compared to a taboo search not employing this mechanism (Fig. 9.3), the taboo duration can be reduced and the search achieves good solutions more frequently. This mechanism could even be operated alone, without a taboo list. Indeed, an excellent solution is obtained with a minimum and maximum taboo duration of 0. A taboo list can be implemented by means of a matrix whose entry (i, j) gives the iteration number from which one can again use the edge $[i, j]$ in a move. Counting the frequency of moves is implemented in a similar way.

9.2.1.3 Restarts

A frequently used technique to intensify a taboo search is to restart with the best solution achieved so far. This is done if the search seems to stagnate, for instance, if there has been no improvement in the best solution during a relatively significant number of iterations. When restarting, the information collected during the previous iterations by the taboo list is kept, as well as other statistics, if any. Hence, the work achieved during these iterations is exploited.

Thus, the data structures guiding the search being in an altered state after restarting, the trajectory followed by the search, will also be. This mechanism can be identified as the opposite of the one presented above where we force the use of neglected attributes for many iterations. Its purpose is to achieve search intensification, not diversification. Naturally, the implementation of this mechanism implies the introduction of new parameters that must be adjusted, like the number of iterations to be carried out before a restart and a possible adaptation of the value of other parameters (taboo duration, frequency penalty) to guide the search toward diversified trajectories.

Problems

9.1 Taboo Search for an Explicit Function

An integer function of integer variables $f(x, y)$ is explicitly given in Table 5.1. We seek the minimum of this function by applying a taboo search. The neighborhood consists in modifying by a unit the value of one variable. The taboo conditions consist in forbidding to increment (respectively: to decrement) a variable that has been decremented (respectively: incremented). First, consider a taboo duration of $d = 3$ and $(-7, -6)$ as the starting solution. Next, start from $(-7, 7)$ and use $d = 1$. The search stops if there is no more move allowed or if 25 iterations have been performed.

9.2 Taboo Search for the VRP

For the VRP, the neighborhood consists in either moving a customer from one route to another, or swapping two customers from different routes. Suggest taboo criteria for this neighborhood.

9.3 Taboo Search for the QAP

Consider the QAP instance given by the flow F and distance D matrices:

$$F = \begin{bmatrix} 0 & 5 & 2 & 4 & 1 \\ 5 & 0 & 3 & 0 & 2 \\ 2 & 3 & 0 & 0 & 0 \\ 4 & 0 & 0 & 0 & 5 \\ 1 & 2 & 0 & 5 & 0 \end{bmatrix} \qquad D = \begin{bmatrix} 0 & 1 & 1 & 2 & 3 \\ 1 & 0 & 2 & 1 & 2 \\ 1 & 2 & 0 & 1 & 2 \\ 2 & 1 & 1 & 0 & 1 \\ 3 & 2 & 2 & 1 & 0 \end{bmatrix}$$

Starting with the solution $p = (1, 2, 3, 4, 5)$, perform six iterations of a taboo search. The moves are defined by pairs (i, j) that swap the elements p_i and p_j. If the move (i, j) is performed, then, it is forbidden for $d = 5$ iterations to place the element p_i in position i and, simultaneously, the element p_j in position j. For each iteration, provide the solution, its value, that of all the moves and their taboo status.

References

1. Battiti, R., Tecchiolli, G.: The reactive tabu search. ORSA J. Comput. **6**, 126–140 (1994). https://doi.org/10.1287/ijoc.6.2.126
2. Glover, F.: Future paths for integer programming and links to artificial intelligence. Comput. Oper. Res. **13**(5), 533–549 (1986). https://doi.org/10.1016/0305-0548(86)90048-1

3. Glover, F.: Tabu search—part I. ORSA J. Comput. **1**(3), 190–206 (1989). https://doi.org/10.1287/ijoc.1.3.190
4. Glover, F.: Tabu search—part II. ORSA J. Comput. **2**(1), 4–32 (1990). https://doi.org/10.1287/ijoc.2.1.4
5. Glover, F., Laguna, M.: Tabu Search. Kluwer, Dordrecht (1997)
6. Taillard, É.D.: Comparison of iterative searches for the quadratic assignment problem. Location Science **3**(2), 87–105 (1995). https://doi.org/10.1016/0966-8349(95)00008-6

Chapter 10
Population Management

By abuse of language, all the methods previously presented can be classified as single-solution metaheuristics. Although most of these methods are building or modifying a lot of different solutions, they only consider one current solution at an iteration and, eventually, the best solution found so far. This classification could be disputed, especially for the ant system. Indeed, several solutions are built at a given iteration. However, an ant constructs a solution without taking care of the work done in parallel by the other ants, and all the solutions built in one iteration are forgotten once the trails are updated. Similarly, there are taboo searches storing several solutions, but they are used for determining the taboo status of a current solution neighbor. This chapter considers methods where several solutions are explicitly stored and iteratively used for generating or modifying other ones.

With a proper modeling of an optimization problem, it is very easy to construct many different solutions, especially by means of a randomized method. Therefore, one can try to learn how to create new solutions from those previously constructed. This chapter studies how to exploit a population of solutions and how to combine the various basic metaheuristic components studied above.

Let us illustrate this by the tour merging technique for the TSP. Figure 10.1 shows five tours obtained with a randomized method in $O(n \log n)$ presented in Section 6.3.2. None of these solutions looks really nice. However, superimposing these solutions on the optimal solution reveals that all the edges of the latter are part of these tours. Therefore, we believe that intelligent exploitation of various solutions can help to discover better ones.

10.1 Evolutionary Algorithms Framework

The intuition at the source of evolutionary algorithms comes from biologist works of the nineteenth century, like Darwin and Mendel who founded the theory of the

É. D. Taillard, *Design of Heuristic Algorithms for Hard Optimization*, Graduate Texts in Operations Research, https://doi.org/10.1007/978-3-031-13714-3_10

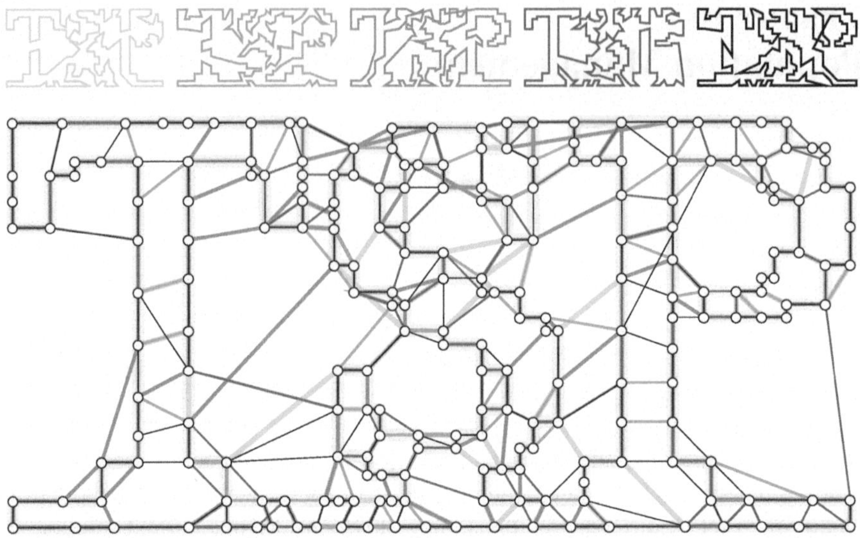

Fig. 10.1 Optimal tour of the TSP instance *tsp225* on which is superimposed five tours obtained by the fast method presented in Section 6.3.2

evolution of living species. Indeed, over the course of generations, living beings are able to adapt to constantly changing external conditions. They can optimize their survival probability, thus resolving extremely complex problems. Therefore, why not attempt to artificially reproduce this evolution to solve hard combinatorial optimization problems?

In the 1960s and 1970s, various ways of exploiting these ideas emerged. The general framework of the evolutionary algorithms is provided by Algorithm 10.1. One begins by generating a set of μ solutions to the problem, usually in a purely random fashion. This set of solutions is called a *population* by analogy with a group of living beings. In the same way, a solution to the problem is an *individual.* Evolutionary algorithms repeat the next loop (called a *generational loop*) until a stopping criterion is met. This is either set in advance, for example, the number of times the generational loop is repeated, or decided on the basis of the diversity of individuals present in the population.

First, a number of solutions from the population are selected to be used for breeding. This is achieved by a *selection operator for reproduction.* The purpose of this operator is to favour the individuals that are well adapted to their environment (those with the best fitness function) at the expense of those that are weaker, sick, and ill-adapted similar to what happens in nature.

The selected individuals are then mixed together (e.g., in pairs) using a *crossover* operator to form λ new solutions called *offspring* which undergo random modifications by means of a *mutation operator.* These two operators simulate the sexual reproduction of living species, assuming that, with a little luck, the favorable characteristics (the desirable genes contained in the DNA) of the parent solutions

---

Algorithm 10.1: Framework of evolutionary algorithms

Input: Parameters μ and λ, selection for reproduction, crossover, mutation and selection for survival operators

Result: Population of solutions P

Generate a population P of μ solutions

repeat

  Select individuals from P with the selection for reproduction operator

  Combine the selected individuals with the crossover operator and apply the mutation operator to get λ new solutions

  Among the $\mu + \lambda$ solutions, select μ individuals with the selection for survival operator; these μ individuals constitute the population P for the next generation

until *a stopping criterion is satisfied*

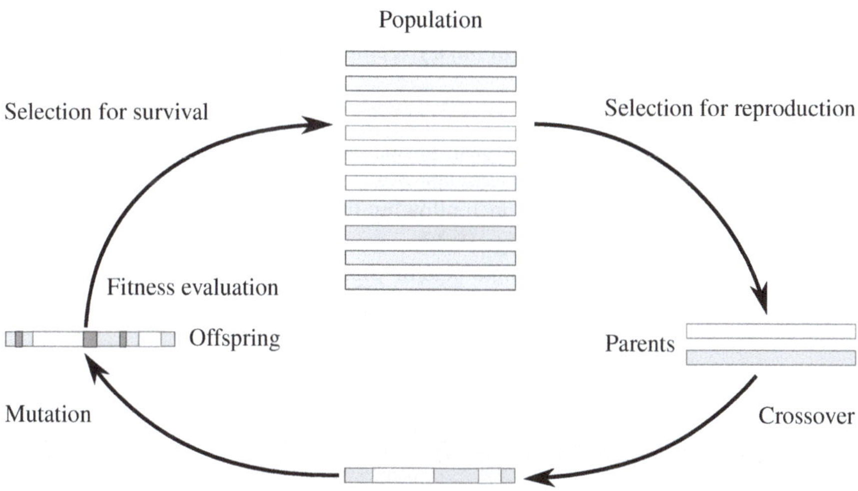

Fig. 10.2 Generational loop in an evolutionary algorithm. From a population of solutions, symbolized here by colored sticks, one selects individuals who reproduce by crossover and mutation. The offspring thus generated is evaluated and incorporated into the population. Ultimately, individuals are eliminated by a selection operator for survival to bring the population back to its initial size

will be transmitted to their children and that fortuitous mutations will result in the appearance new favorable genes.

Finally, the new solutions are evaluated, and a *selection operator for survival* eliminates λ solutions from the $\mu + \lambda$ available to reduce to a new population of μ individuals. Figure 10.2 illustrates the process of a generational loop.

The framework of evolutionary algorithms leaves considerable freedom in the choices to be made for the implementation of the various operators and parameters. For instance, the "evolution strategy" of Rechengerg [9] does not use a crossover operator between two solutions. In this technique, the solutions of the population are modified with a mutation operator and compete with each other, much like parthenogenetic reproduction.

10.2 Genetic Algorithms

Among evolutionary algorithms, it is undoubtedly the genetic algorithms (GA) proposed by Holland [3] that have received the most attention. This is paradoxical, since the purpose of his study was to understand the convergence mechanisms of these algorithms, not their ability to optimize difficult problems. For a long time, the community in this field continued to work on the genetic algorithm convergence theory, studying "orthodox" versions of the various operators mentioned above, in conjunction with a standard representation of solutions under the form of Boolean vectors with a specified size.

Unfortunately, not all optimization problems have solutions that can be naturally represented by binary vectors. Using only standard operators and knowing their theoretical properties, considerable efforts have been made to discover appropriate encodings of solutions in the form of binary vectors and to decode them into feasible solutions.

For the problems whose solutions are naturally represented by a permutation, the *random key* coding technique allows exploiting the standard crossover and mutation operators. A permutation of the elements of $1 \dots n$ are represented by an array t of n real numbers. The permutation p allowing the sorting of t corresponds to the solution coded by the array (see Fig. 10.12).

The next sections review the main genetic algorithm operators, discussing how they can be generalized so that they equally apply to a natural representation of solutions and not only to binary vectors.

10.2.1 Selection for Reproduction

The selection for reproduction aims to favor the most efficient solutions so that they can transmit their beneficial properties to their offspring. Each solution i must therefore be assigned a fitness measure f_i; the higher the quality, the higher the selection probability must be. If the objective of the problem to be solved is to maximize a function admitting positive values, this function can be directly used as fitness function. Otherwise, a transformation of the objective function is required to assign a fitness to each individual.

10.2.1.1 Rank-Based Selection

A traditional transformation is to sort the individuals. This does not require the computation of an objective function but only the possibility to compare the solution quality. The fittest individual in a population has a rank of 1 and the worst of μ.

The individual i of rank r_i has a quality measure $f_i = (1-\frac{r_i-1}{\mu})^p$, where $p \geqslant 0$ is a parameter to modulate the selection pressure. A pressure $p = 0$ implies a uniform draw among the population (no selective pressure), while $p = 2$ represents a fairly high pressure. Code 10.1 provides an implementation of this operator for a selection pressure of $p = 1$.

Code 10.1 rank_based_selection.py Implementation of a rank-based selection operator for reproduction, with selective pressure $p = 1$. The best of μ individuals has a probability of $\frac{2\mu}{\mu\cdot(\mu+1)}$ to be selected, while the worst has a probability of $\frac{2}{\mu\cdot(\mu+1)}$

```
import math
from random_generators import unif                    # Listing 12.1

######### Selection operator for reproduction based on the rank
def rank_based_selection(size):
    return int(size \
            - math.ceil(math.sqrt(.25 + 2*unif(1, size*(size + 1)/2)) - .5))
```

10.2.1.2 Proportional Selection

The simplest selection operator is to randomly draw an individual proportionally to its fitness. The individual i has thus a probability $f_i/\sum f_i$ of being selected. In principle, we do not select just one individual at each generational loop but several. The selection is ordinarily performed with replacement, so that a (good) individual can be selected several times in one generation.

Genetic algorithms are inherently parallel: the generational loop can be applied both to the production of a unique individual in each generation, as shown in Fig. 10.2, and to the generation of a multitude of offspring. A frequently used technique is to select an even number λ of parent solutions in a generation and pair them up, and each pair produces two offspring per crossover.

10.2.1.3 Natural Selection

It is also possible to perform a purely random and uniform selection for reproduction, just like what happens to many living species. The convergence of the algorithm must then be guided by the selection operator for survival, which ensures a bias toward the fittest individuals. Table 10.1 compares the selection probabilities of the operators presented above for a small population.

10.2.1.4 Complete Selection

If one does not choose too large a population size, it is also possible to involve all individuals in a systematic way for reproduction. As with natural selection, the

Table 10.1 Selection probability for different operators for reproduction. The objective function is to be maximized and is directly used as a fitness function for the proportional selection. The sum of the values of the objective function is 1000

Objective function	Rank	Probability Rank-based ($p = 2$)	Rank-Based ($p = 1$)	Natural	Proportional
220	1	0.260	0.182	0.1	0.220
162	2	0.210	0.164	0.1	0.162
157	3	0.166	0.146	0.1	0.157
93	4	0.127	0.127	0.1	0.093
85	5	0.094	0.109	0.1	0.085
74	6	0.065	0.091	0.1	0.074
61	7	0.042	0.073	0.1	0.061
55	8	0.023	0.054	0.1	0.055
49	9	0.010	0.036	0.1	0.049
44	10	0.003	0.018	0.1	0.044

evolution of the population toward good solutions then depends on the selection operator for survival, which should favor the best solutions.

10.2.2 Crossover Operator

A crossover operator aims to simulate the sexual reproduction of living species. Schematically, the process of meiosis in sexual reproduction separates the DNA of each parent into two genetic sequences. This produces gametes (egg cell, sperm, or pollen grains). During the fertilization of the egg cell, genetic shuffling occurs, during which the sequence of genes of the offspring is produced by sequentially adding the genes of either parent in an arbitrary fashion.

The purpose of this operator is to produce a new offspring, different from its parents, but having inherited some of their features. With a little luck, the offspring receives good features from its parents and is better adapted to its environment. With a little less luck, the offspring does not receive those good features. Nevertheless, it perpetuates valuable genes and provides a source of diversity within the population, which means potential for innovation.

Figure 10.3 metaphorically illustrates this with the mating of different ladybird beetles. The couples at the top are likely to produce children very similar to themselves, while the couples at the bottom of the figure might produce genetically richer children.

There are evolutionary strategies where the crossover operator is absent. These strategies mimic asexual reproduction, where an individual produces an offspring practically identical to itself, where only spontaneous mutations cause the population gene pool to evolve.

Fig. 10.3 Ladybird beetles mating. One can imagine the top couples will produce children very similar to themselves, while the lower ones will keep some genetic diversity in the population and, hopefully, produce some children better adapted to their environment

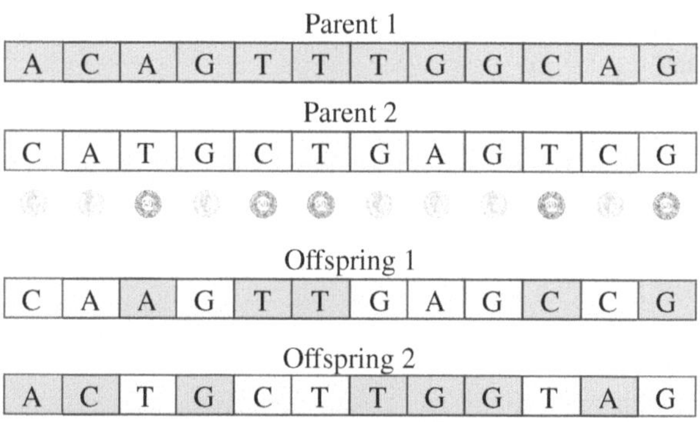

Fig. 10.4 Uniform crossover. Production of two complementary offspring from the genes of two parents. Each item of the first offspring is chosen at random from either parent by flipping a coin. The second offspring receives the complementary item

10.2.2.1 Uniform Crossover

Uniform crossover involves taking two parent solutions, represented as vectors of n items and creating a third one by choosing its items from either parent with equal probability. Figure 10.4 illustrates the production of two "anti-twin" offspring from two parents. This crossover operator is appropriate if it is straightforward and logical to represent any solution of the problem by a vector of n components and if any vector of that size can match a feasible solution.

This is not the case for a problem where a permutation of n items is sought. One technique for adapting the uniform crossover for this situation is to proceed in two phases: in the first phase, the items of the permutation are randomly selected from either parent, provided that the item has not yet been chosen. If both parents possess items already selected at the position to be filled, the latter remains temporarily

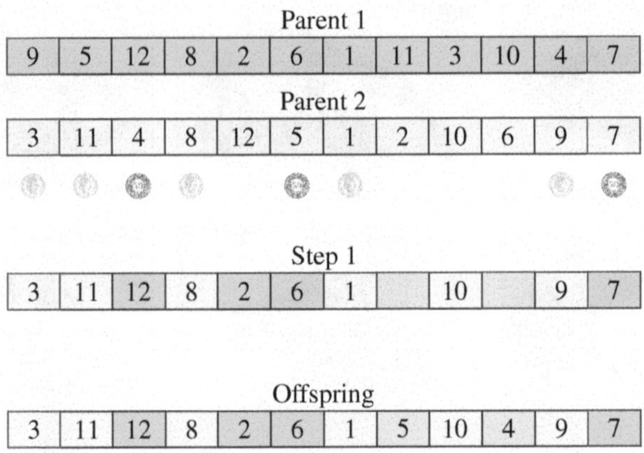

Fig. 10.5 Uniform crossover on a permutation. An offspring is produced in two phases. We first successively choose the items of one or the other of the parents, as long as they are not part of the offspring. Otherwise, either we leave the position empty if both items are already in the offspring or we select the unique item available. The second phase randomly completes the offspring using the remaining items

empty in the offspring. The second phase consists in filling in at random the vacant positions with the items that were not selected during the first phase. This operator is illustrated in Fig. 10.5.

10.2.2.2 Single-Point Crossover

The single-point crossover first randomly picks a point within the solution vector. Then it copies all the items of the first parent up to that point. Finally, it copies the items of the second parent from there. In practice, for a vector of n items, we randomly draw a number c between 1 and $n - 1$; we copy the items 1 to c from the first parent and the items $c + 1$ to n from the second parent. We can produce a second complementary offspring in parallel. Figure 10.6 illustrates this operator.

10.2.2.3 Two-Point Crossover

The two-point crossover consists in randomly selecting two different points. The offspring is created by copying the items before the first point and after the second point from one parent and copying the portion between the points from the other parent. This operator is illustrated in Fig. 10.7. The strategy can be generalized by choosing k crossover points.

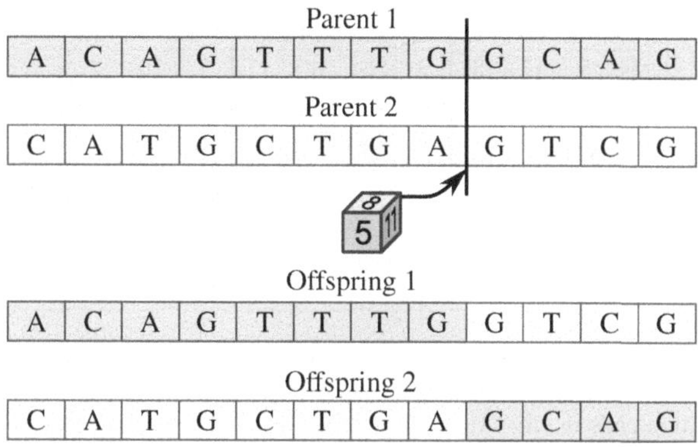

Fig. 10.6 Single-point crossover. Production of two "anti-twins" by randomly drawing a crossover point (here, the 8). The items of the first parent are copied up to the crossover point and those of the second from there on

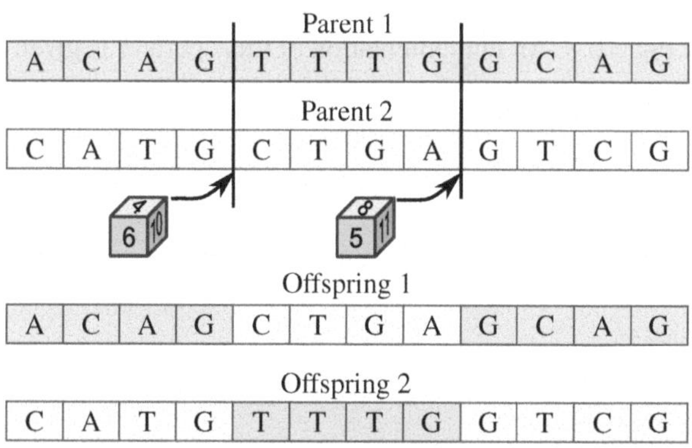

Fig. 10.7 Two-point crossover. Production of two "anti-twins" by randomly drawing two crossover points (here, the 4 and 8). The items of the first parent are copied up to the first and from the second crossover point. The intermediate items come from the second parent

10.2.2.4 OX Crossover

For each problem, we can invent a specific crossover operator. For instance, for the TSP, one can advance the argument that portions of the paths should be copied from the parents into the offspring. If a solution is a permutation of the cities, we realize that the uniform crossover seen previously (adapted to the case of permutations) does not really make sense: the starting city is not decisive. The cities that precede and succeed a given city are important, not the absolute position of the city in the tour. The two-point crossover operator can be adapted for the problems where the sequences are significant.

The OX crossover operator devised for the TSP begins by copying the intermediate portion of one parent, like the two-point crossover. The last city of this portion is located in the other parent and the offspring is completed by cyclically scanning the

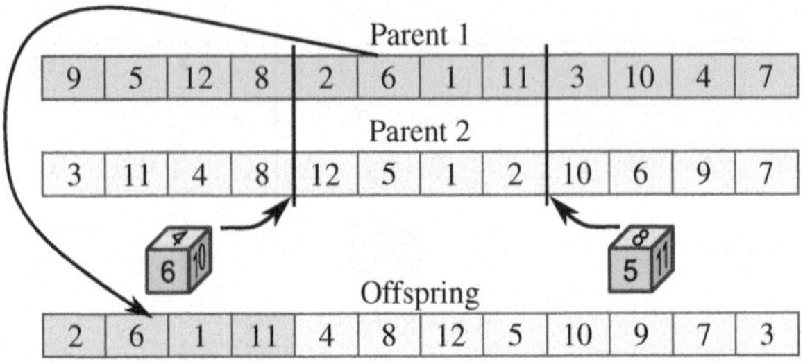

Fig. 10.8 OX crossover, specifically developed for the TSP. We start by randomly drawing two crossover points. The intermediate portion of the first parent is copied to the offspring. In this example, this portion ends in city 11. We locate this city in the second parent and complete the offspring from there, in this case with city 4. The cities already appearing in the offspring (1, 2, and 6) are skipped. When we arrive at the last city of the second parent (7), we return to the first (3) one

cities of this parent and inserting those not yet included. The OX crossover operator is illustrated in Fig. 10.8. An implementation of this operator is given in Code 10.2.

Code 10.2 **OX_crossover.py** Implementation of the OX crossover operator, preserving a sub-path

```
from random_generators import unif                              # Listing 12.1

######### Crossover operator preserving successive values in a permutation
def OX_crossover(parent1, parent2):                             # Parent solutions

    n = len(parent1)
    # Randomly generate the portion of parent1 that is copied in child
    point2, point1 = unif(1, n - 2), unif(1, n - 3)
    if point1 >= point2:
        temp = point2
        point2 = point1 + 1
        point1 = temp

    # Copy the portion of parent1 at the beginning of child
    child = [-1] * n
    inserted = [0] * n              # Flag for elements already inserted in child
    for i in range(point2 - point1 + 1):
        child[i] = parent1[i + point1]
        inserted[child[i]] = 1

    # Last element of parent2 inserted in child
    i = parent2.index(child[point2 - point1])

    # Insert remaining elements in child, in order of appearance in parent2
    nr_inserted = point2 - point1 + 1
    while nr_inserted < n:
        if not inserted[parent2[i % n]]:
            child[nr_inserted] = parent2[i % n]
            inserted[parent2[i % n]] = 1
            nr_inserted += 1
        i += 1
    return child
```

10.2.3 Mutation Operator

The mutation operators can be described in a simple way in the context of this book: it consists in randomly applying one or more local moves to the solution, as described in Chap. 5 devoted to local searches.

The mutation operator has two roles: firstly, the local modification can improve the solution, and, secondly, even if the solution is not improved, it slows down the global convergence of the algorithm by strengthening the genetic diversity of the population. Indeed, without this operator, the population can only lose diversity. For instance, the crossover operators presented above systematically copy the identical parts of the parents in the offspring. Thus, some genes take over compared to others that disappear with the elimination of solutions by the selection operator for survival.

Figure 10.9 illustrates the influence of the mutation rate for a problem where a permutation of n elements is sought. In this figure, a mutation rate of 5% means that there is such a proportion of elements that are randomly swapped in

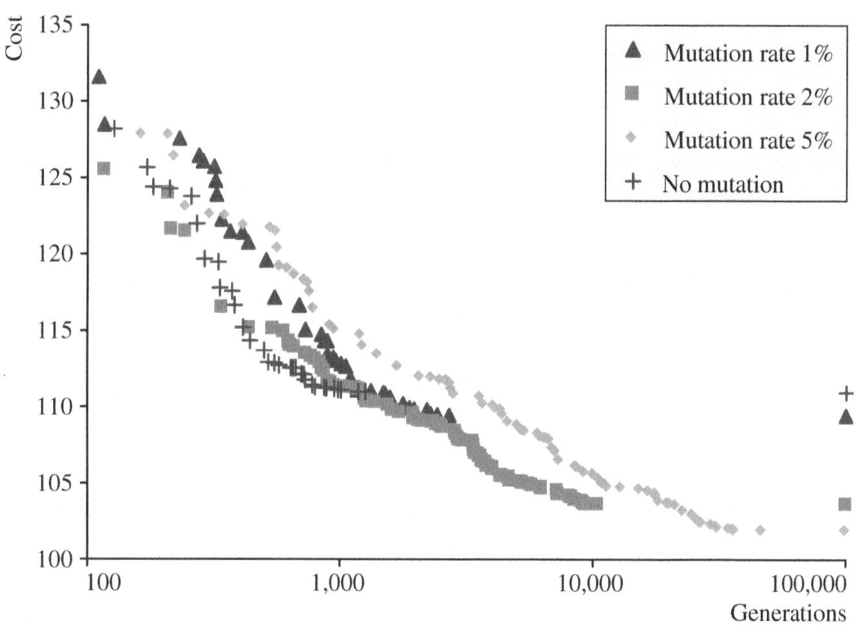

Fig. 10.9 Influence of the mutation rate on the quality of the solutions produced as a function of the number of generational loops performed. Only the value of the best solution in the population is indicated. The algorithm favors the best solutions in the population by means of selection operators for reproduction and survival. Without mutation, the population converges relatively rapidly to individuals that are all similar and of poor quality. The higher the mutation rate, the slower the convergence, resulting in better solutions

the permutation. Code 10.3 gives an implementation of a mutation operator for problems on permutations.

Code 10.3 mutate.py Implementing a mutation operator for problems on permutations

```
from random_generators import unif                           # Listing 12.1

######### Random mutation of a permutation
def mutate(mutation_rate,
           p):                                              # Permutation to mutate

    n = len(p)
    mutations = int(mutation_rate * n / 2.0)
    for _ in range(mutations):
        i = unif(0, n - 1)
        j = unif(0, n - 1)
        p[i], p[j] = p[j], p[i]
    return p
```

10.2.4 Selection for Survival

The last key operator in genetic algorithms is selection for survival, which aims to bring the population back to its initial size of μ individuals, after λ new solutions have been generated. Several selection policies have been devised, depending on the values chosen for the parameters μ and λ.

10.2.4.1 Generational Replacement

The simplest policy for selecting the individuals who will survive is to generate the same number of offspring as there are individuals in the population ($\lambda = \mu$). The population at the beginning of the new generational loop is made up only of the offspring, the initial population disappearing. With such a choice, it is necessary to have a selection operator for reproduction that favors the best solutions. This means the best individuals are able to participate in the creation of several offspring, while some of the worst are excluded from the reproduction process.

10.2.4.2 Evolutionary Strategy

The evolutionary strategy (μ, λ) consists in generating numerous offspring ($\lambda > \mu$) and in only keeping the μ best offspring for the next generation. The population is therefore completely changed from one iteration of the generational loop to the next. This strategy leads to a bias in the choice of the fittest individuals from one generation to the next. So, it is compatible with a uniform selection operator for reproduction.

10.2.4.3 Stationary Replacement

Another commonly used technique is to gradually evolve the population, with the generation of few offspring at each generational loop. A strategy is to generate $\lambda = 2$ children in each generation, which will replace their parents.

10.2.4.4 Elitist Replacement

Another more aggressive strategy is to consider all the $\mu + \lambda$ solutions available at the end of the generational loop and to keep only the best μ for the next generation. This strategy was adopted to produce Fig. 10.10 illustrating the evolution of the fittest solution of the populations for various values of μ.

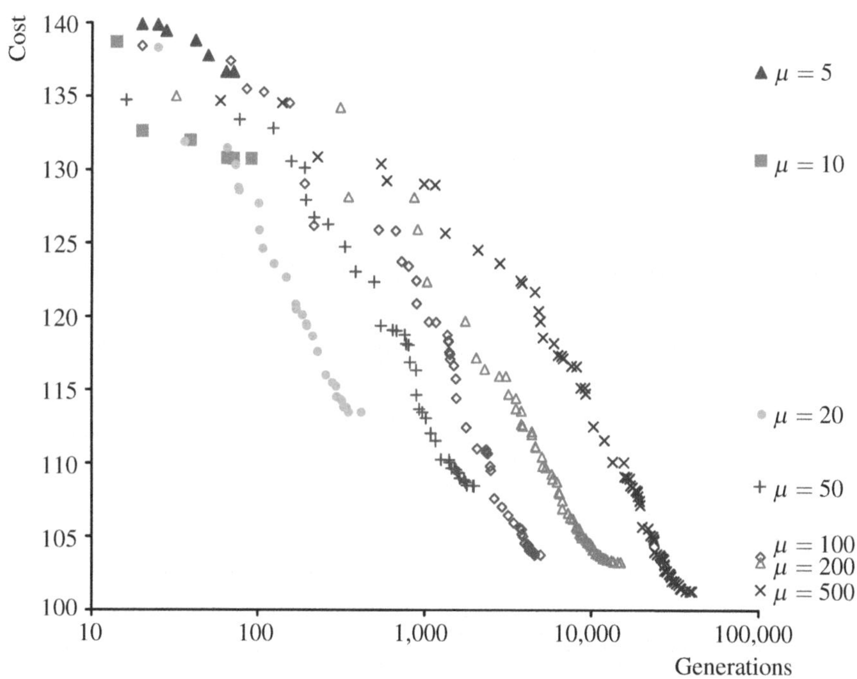

Fig. 10.10 Influence of the population size on the solution quality. When the population is too limited, it converges very rapidly with a low probability of discovering good solutions. Conversely, a large population converges very gradually, but better solutions are obtained

Code 10.4 implements an elitist replacement when $\lambda = 1$, which means that only one offspring is produced at each generation. It replaces the worst solution in the population (if it is not even worse). In this code, we have included a basic population management that must contain exclusively different individuals. To simplify the test of equality between two solutions, they are discriminated only on the basis of their fitness: two solutions of the same length are considered identical.

Code 10.4 insert_child.py Implementation of elitist replacement where each generation produces only one child. This procedure implements basic population management where all individuals must have different fitness

```
######### Inserting a child in a population of solutions
def insert_child(child,                        # Individual to insert in population
                 child_fitness,       # Cost of child (the smaller, the better)
                 population_size,
                 population,
                 fitness,                             # Fitness of each individual
                 order):                   # order[i] : individual number with rank i

    rank = [-1 for _ in range(population_size)]             # Rank of individuals
    for i in range(population_size):
        rank[order[i]] = i

    child_rank = 0                                        # Find the rank of the child
    for i in range(population_size):
        if fitness[i] < child_fitness:
            child_rank += 1

    if child_rank < population_size - 1:              # The child is not dead-born
        if fitness[order[child_rank]] != child_fitness \
           and (child_rank == 0
                or fitness[order[child_rank - 1]] != child_fitness):
            population[order[population_size - 1]] = child[:]
            fitness[order[population_size - 1]] = child_fitness

            for i in range(population_size):
                if rank[i] >= child_rank:
                    rank[i] += 1
            rank[order[population_size - 1]] = child_rank

            for i in range(population_size):
                order[rank[i]] = i
        else:
            child_rank = population_size
    return child_rank, population, fitness, order
```

10.3 Memetic Algorithms

Genetic algorithms have two major drawbacks: first, nothing ensures that the best solution found cannot be improved by a simple local modification, as seen in Chap. 5. Second, the diversity of the population declines with each iteration of the generational loop, eventually consisting only of clones of the same individual.

To overcome these two drawbacks, Moscato [8] designed what he called *memetic algorithms*. The first of these shortcomings is solved by applying a local search after producing an offspring. The simplest way to avoid duplication of individuals in the population is to eliminate them immediately, as implemented in Code 10.4.

Code 10.5 illustrates a straightforward implementation of a memetic algorithm for the TSP where the offspring are improved using a local search based on ejection chains and only replace the worst solution in the population if they are of better quality than the latter and their evaluation is different from all those in the population, thus ensuring that no duplicates are created. This algorithm implements only an elementary version of a memetic algorithm.

Code 10.5 tsp_GA.py Implementation of a memetic algorithm for the TSP. This algorithm uses a selection operator for reproduction based on rank. After its generation, the offspring is improved by a local search (ejection chain method) and immediately replaces the worst solution in the population. This algorithm has three parameters: the number μ of solutions in the population, the number of generational loops to be performed, and the mutation rate

```
from random_generators import rand_permutation                      # Listing 12.1
from tsp_utilities import tsp_length                                # Listing 12.2
from rank_based_selection import *                                  # Listing 10.1
from OX_crossover import OX_crossover                               # Listing 10.2
from mutate import mutate                                           # Listing 10.3
from insert_child import insert_child                               # Listing 10.4
from tsp_LK import tsp_LK                                           # Listing 12.3

######### Basic Memetic Algorithm for the TSP
def tsp_GA(d,                                # Distance matrix (must be symmetrical)
           population_size,                                  # Size of the population
           generations,                                      # Number of generations
           mutation_rate):

    n = len(d[0])
    population = [rand_permutation(n) for _ in range(population_size)]
    lengths = [tsp_length(d, population[i]) for i in range(population_size)]

    order = [i for i in range(population_size)]
    for i in range(population_size - 1):
        for j in range(i + 1, population_size):
            if lengths[order[i]] > lengths[order[j]]:
                order[i], order[j] = order[j], order[i]
    print('GA initial best individual {:d}'.format(lengths[order[0]]))

    for gen in range(generations):
        parent1 = rank_based_selection(population_size)
        parent2 = rank_based_selection(population_size)
        child = OX_crossover(population[order[parent1]],
                             population[order[parent2]])
        child = mutate(mutation_rate, child)
        child_length = tsp_length(d, child)
        child, child_length = tsp_LK(d, child, child_length)
        child_rank, population, lengths, order = insert_child(child,
                    child_length, population_size, population, lengths, order)
        if child_rank == 0:
           print('GA improved tour {:d} {:d}'.format(gen, child_length))
    return population[order[0]], lengths[order[0]]
```

Sörensen and Seveaux [11] proposed a more advanced population management. These authors suggest evaluating, for each solution produced, a similarity measure with the solutions contained in the population. Solutions that are too similar are discarded to maintain sufficient diversity so that the algorithm does not converge prematurely.

10.4 Scatter Search

Scatter search is almost as old as genetic algorithms. Glover [1] proposed this technique in the context of integer linear programming. At the time, it broke certain taboos, such as being able to represent a solution in a natural form and not coded by a binary vector or to mix more than two solutions between them, as metaphorically illustrated in Fig. 10.11.

The chief ideas of scatter search comprise the following characteristics, presented in contrast to traditional genetic algorithms:

Dispersed initial population Rather than randomly generating a large initial population, the last is generated deterministically and as scattered as possible in the space of potential solutions. They are not necessarily feasible but are rendered so by a repair/improvement operator.

Natural representation of solutions Solutions are represented in a natural way and not necessarily with binary vectors of a given size.

Combination of several solutions More than two solutions may contribute to the production of a new potential solution. Rather than relying on a large population and a selection operator for reproduction, scatter search tries all possible combinations of individuals in the population, which must therefore be limited to a few dozen solutions.

Repair/improvement operator Because of the natural representation of solutions, the simultaneous combination of several individuals does not necessarily produce a feasible solution. A repair operator projecting a potential infeasible solution

Fig. 10.11 Scatter research breaks the taboo of breeding limited to two solutions

into the space of feasible solutions is therefore expected. This operator can also improve a feasible solution, especially by means of a local search.

Population management A reference population, of small size, is decomposed into a subset of *elite solutions* (the best ones) and other solutions as different as possible from the elites. The goal is to increase the diversity of the population while keeping the best solutions.

The framework of scatter search is given by Algorithm 10.2. The advantage of this framework is its limited number of parameters: μ for the size of the reference population and $E < \mu$ for the set of elite solutions. Moreover, the value of μ must be limited to about twenty, since it is necessary to combine a number of potential solutions increasing exponentially with μ; this also means that the number E of elite solutions should be from a few units to about ten.

Algorithm 10.2: Scatter search framework

Input: Size μ of the complete population, E size of the subset of elite solutions
Result: Population of solutions

Systematically generate a (large) population P of potential solutions as dispersed as possible
repeat
  Repair and improve the solutions from P to make them feasible using the repair/improvement operator
  Eliminate identical solutions from P
  Identify the E best solutions from the population; they are retained in the reference set as elites
  Identify from P the $\mu - E$ solutions which are the most different from the elite solutions, they are kept and complete the reference set
  Combine in all possible ways the μ solutions of the reference set to obtain $2^{\mu} - \mu - 1$ new potential solutions
  Join the potential solutions to the reference set to obtain the new population P of the next iteration
until *the population remains stable*

10.4.1 Illustration of Scatter Search for the Knapsack Problem

To illustrate how the various options in the scatter search framework can be adapted to a particular problem, let us consider a knapsack instance:

$$
\begin{aligned}
&\max r = 11s_1 + 10s_2 + 9s_3 + 12s_4 + 10s_5 + 6s_6 + 7s_7 + 5s_8 + 3s_9 + 8s_{10}\\
&\text{Subject } 33s_1 + 27s_2 + 16s_3 + 14s_4 + 29s_5 + 30s_6 + 31s_7 + 33s_8 + 14s_9 + 18s_{10} \leqslant 100\\
&\quad\text{to: } s_i \in \{0, 1\} (i = 1, \dots, 10)
\end{aligned}
\tag{10.1}
$$

10.4.1.1 Initial Population

The solutions to this problem are, therefore, ten-component binary vectors. To generate a set of potential solutions as scattered as possible, one can choose to put either all the objects in the knapsack, or one out of two, or one out of three, etc. For each potential solution thus generated, the complementary solution can also be added to the population. Naturally, not all the solutions from the population are feasible. To be specific, the solution with all objects does not satisfy the knapsack volume constraint; its complementary solution, with an objective value of zero, is the worst possible.

A repair/improvement operator must therefore be applied to these potential solutions. This can be performed as follows: as long as the solution is not feasible, remove the object with the worst value/volume ratio. A feasible solution can be improved greedily, by including the object with the best value/volume ratio as long as the capacity of the knapsack permits it. This produces the population of solutions given in Table 10.2.

10.4.1.2 Creation of the Reference Set

Solutions 9 and 10 are identical to the first solution and are therefore eliminated. If we choose a set of $E = 3$ elite solutions, these are solutions 1, 2, and 8. Assuming that one wishes a reference set of $\mu = 5$ solutions, two solutions must be added to the three elites, among solutions 3 to 7. The two solutions to complete the reference set are determined by evaluating a measure of dissimilarity with the elites. An approach is to consider the solutions maximizing the smallest Hamming distance to one of the elites which is illustrated in Table 10.3.

Table 10.2 Initial scattered population P for the knapsack instance 10.1 and the result of applying the repair/improvement operator on the potential solutions. Those which are not feasible are in bold, as well as the $E = 3$ elite solutions

	Potential solution	Value	Repaired/improved solution	Value	
1	**(1,1,1,1,1,1,1,1,1,1)**	81	**(0,1,1,1,0,0,0,0,1,1)**	42	
2	**(1,0,1,0,1,0,1,0,1,0)**	40	**(1,0,1,1,1,0,0,0,0,0)**	42	
3	(1,0,0,1,0,0,1,0,0,1)	38	(1,0,0,1,0,0,1,0,0,1)	38	
4	(1,0,0,0,1,0,0,0,1,0)	24	(1,0,0,1,1,0,0,0,1,0)	36	
5	(1,0,0,0,0,1,0,0,0,0)	17	(1,0,1,1,0,1,0,0,0,0)	38	
6	(0,0,0,0,0,0,0,0,0,0)	0	(0,1,1,1,0,0,0,0,0,1)	39	
7	**(0,1,0,1,0,1,0,1,0,1)**	41	(0,1,0,1,0,1,0,0,0,1)	36	
8	**(0,1,1,0,1,1,0,1,1,0)**	43	**(0,1,1,1,1,0,0,0,1,0)**	44	
9	**(0,1,1,1,0,1,1,1,0,1)**	57	(0,1,1,1,0,0,0,0,1,1)	42	= solution 1
10	**(0,1,1,1,1,0,1,1,1,1)**	64	(0,1,1,1,0,0,0,0,1,1)	42	= solution 1

Table 10.3 Determining the solutions from the population that are as different as possible from the elites. If we want a reference set of $\mu = 5$ solutions, we retain solutions 3 and 7 in addition to the three elites because they are those maximizing the smallest distance to one of the elites

	Candidate solution	Hamming distance			Minimal distance
		Elite 1	Elite 2	Elite 8	
3	(1,0,0,1,0,0,1,0,0,1)	5	4	7	4
4	(1,0,0,1,1,0,0,0,1,0)	5	2	3	2
5	(1,0,1,1,0,1,0,0,0,0)	5	2	5	2
6	(0,1,1,1,0,0,0,0,0,1)	1	4	3	1
7	(0,1,0,1,0,1,0,0,0,1)	3	6	5	3

10.4.1.3 Combining solutions

Finally, we need to implement an operator that allows us to create a potential solution by combining several of them from the reference set. Let us suppose we want to combine solutions 3, 7, and 8, of values 38, 36, and 44, respectively. One possibility is to consider the solutions as numerical vectors and make a linear combination of them. It is tempting to assign a weight according to the solution's fitness. One idea is to give a weight of $\frac{38}{38+36+44}$ to solution 3, of $\frac{36}{38+36+44}$ to solution 7, and of $\frac{44}{38+36+44}$ to solution 8. The vector thus obtained is rounded to project it to binary values:

$$\begin{aligned}
&0.322 \cdot (1, 0, 0, 1, 0, 0, 1, 0, 0, 1) + \\
&0.305 \cdot (0, 1, 0, 1, 0, 1, 0, 0, 0, 1) + \\
&0.373 \cdot (0, 1, 1, 1, 1, 0, 0, 0, 1, 0) \\
&\quad = (0.322, 0.678, 0.373, 1.000, 0.373, 0.305, 0.322, 0.000, 0.373, 0.627) \\
&\mathit{Rounded} : (0, 1, 0, 1, 0, 0, 0, 0, 0, 1)
\end{aligned}$$

10.5 Bias Random Key Genetic Algorithm

Biased random key genetic algorithms (BRKGA) also provide population management with a subset of E elite solutions that are copied to the next generation. The main ingredients of this technique are:

- An array of real numbers (keys) encodes a solution. If a natural representation of a solution is a permutation, then the permutation is one that sorts the keys in increasing order.
- The E best solutions from the population are kept for the next generation.
- The selection operator for reproduction always chooses a solution among the E best.

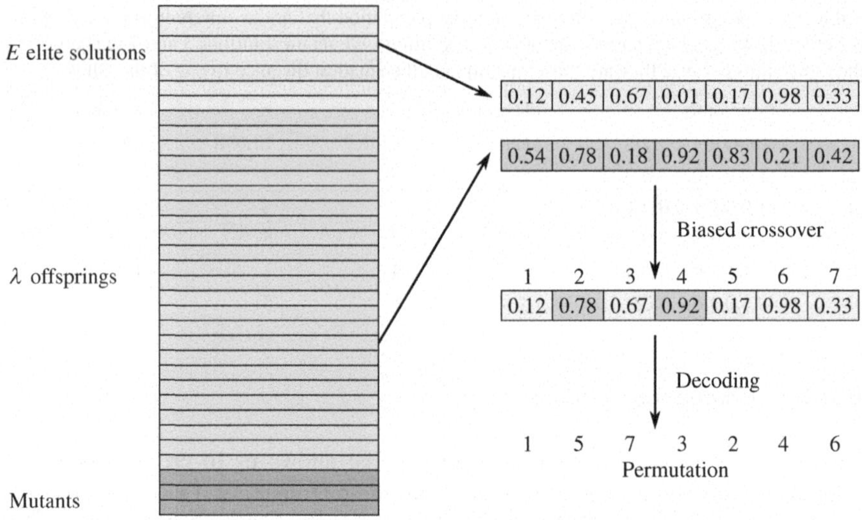

Fig. 10.12 BRKGA: elite solutions are copied from one generation to the next one; a parent always comes from the elite; the crossover operator is biased and chooses more elements from the best parent; the offspring is decoded by sorting the keys in increasing order; the order provides the permutation associated with a solution

- An offspring is generated with a uniform crossover operator, but the components of the best parent-solution are chosen with probability $> 1/2$.
- At each generation, $\lambda < \mu - E$ children are generated. These offspring replace non-elite solutions for the next iteration.
- The genetic diversity of the population is ensured by the introduction of $\mu - E - \lambda$ new randomly drawn arrays (mutants); this replaces the mutation operator.

Figure 10.12 illustrates how this method operates to generate a new solution.

10.6 Path Relinking

Path relinking (PR) was proposed by Glover [2] in the context of taboo search. The idea is to memorize a number of good solutions found by a taboo search. We select two of these solutions, which have been linked by a path with the taboo search. We link these two solutions again by a new, shorter path, going from neighboring solution to neighboring solution.

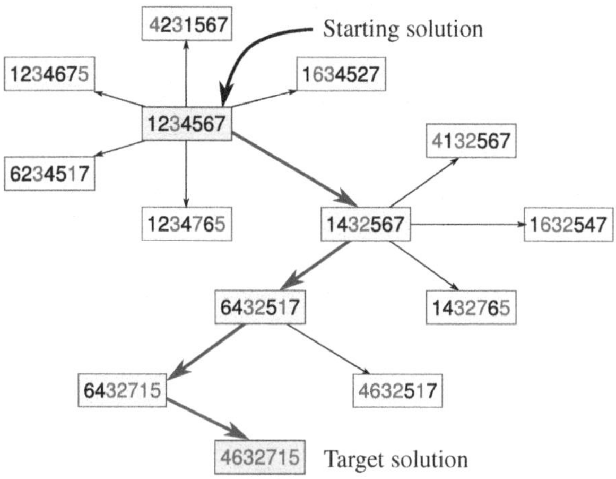

Fig. 10.13 Path relinking. A starting solution—here, the permutation of seven elements (1, 2, 3, 4, 5, 6, 7)—is progressively transformed into a target solution (4, 6, 3, 2, 5, 1, 7) with a neighborhood structure. At each step, the neighbor solutions that are closer to the target solution are evaluated, and the one with the best fitness is chosen

This technique can be implemented independently of a taboo search since all that is needed to implement it is a population of solutions and a neighborhood structure. A starting solution and an ending (target) solution are chosen from the population. We evaluate all the neighbors of the starting solution that are closer to the target solution than the starting one. Among these neighbors, the one with the best evaluation is identified, and the process is repeated from there until we arrive at the target solution. With a bit of luck, one of the intermediate solutions improves the best solution discovered. The path relinking technique is illustrated in Fig. 10.13.

There are different versions of path relinking: the path can be traversed in both directions by reversing the role of the starting and target solutions; an improvement method can be applied to each intermediate solution; ultimately, the starting and target solutions can be alternately modified, and the process stops when meeting in an intermediate solution. Code 10.6 provides an implementation of path relinking for the TSP. It is based on 3-opt moves.

Code 10.6 tsp_path_relinking.py Path relinking implementation for the TSP. At each iteration, we identify a 3-opt move that incorporates at least one arc from the target solution to the current solution

```
from tsp_utilities import tsp_succ_to_pred                          # Listing 12.2

######### Path relinking for the TSP, based on 3-opt neighborhood
def tsp_path_relinking(d,                                            # Distance matrix
                       target,                          # Target solution (successors)
                       length,                            # Length of current solution
                       succ):                                      # Starting solution

    best_succ = succ[:]
    best_length = length
    pred = tsp_succ_to_pred(succ)
    best_delta = -1
    while best_delta < float('inf'):
        best_delta = float('inf')
        i = best_i = best_j = best_k = pred[0]
        while best_delta >= 0 and i != 0:
            i = succ[i]
            if succ[i] != target[i]:
                j = pred[target[i]]
                k = target[i]
                while k != i:
                    if succ[k] != target[k]:
                        delta = d[i][succ[j]] + d[j][succ[k]] + d[k][succ[i]] \
                                -d[i][succ[i]] - d[j][succ[j]] - d[k][succ[k]]
                        if delta < best_delta:
                            delta = best_delta
                            best_i, best_j, best_k = i, j, k
                    k = succ[k]
        if best_delta < float('inf'):
            i, j, k = best_i, best_j, best_k
            length += best_delta;
            pred[succ[i]], pred[succ[j]], pred[succ[k]] = k, i, j;
            succ[j], succ[k], succ[i] = succ[k], succ[i], target[i];
            if length < best_length:
                best_length = length
                best_succ = succ[:]
    return best_succ, best_length
```

10.6.1 GRASP with Path Relinking

A method using the core components of metaheuristics (construction, local search, and management of a population of solutions) while remaining relatively simple and with few parameters is the *GRASP-PR* method (greedy adaptive search procedure with path relinking) by Laguna and Marti [6]. The idea is to generate a population P of different solutions by means of a GRASP with a parameter α (see Algorithm 7.8). These solutions are improved by means of a local search.

Then, we repeat I_{max} times a loop where we build a new solution, greedily and with a bias. This solution is also improved with a local search. We then randomly draw another solution of P and apply a path relinking procedure between both solutions.

The best solution of the path is added to P if it is both strictly better than one of P and is not already present in P. The new solution replaces the solution of P which is the most different from itself while being worse.

Algorithm 10.3 provides the GRASP-PR framework. Code 10.7 implements a GRASP-PR method for the TSP. The reader interested in recent GRASP-based optimization tools can find extensive information in the recent book of Resende and Ribeiro [10].

Algorithm 10.3: GRASP-PR framework

Input: GRASP procedure (with local search `LS` and parameter $0 \leqslant \alpha \leqslant 1$), parameters I_{max} and μ
Result: Population P of solutions

$P \leftarrow \varnothing$
while $|P| < \mu$ **do**
  $s \leftarrow GRASP(\alpha, \texttt{LS})$
  if $s \notin P$ **then**
    $P \leftarrow P \cup s$
for I_{max} *iterations* **do**
  $s \leftarrow GRASP(\alpha, \texttt{LS})$
  Randomly draw $s' \in P$
  Apply a path relinking method between s and s'; identifying the best solution s'' of the path
  if $s'' \notin P$ *and* s'' *is strictly better than a solution of* P **then**
    s'' replaces the most different solution of P which is worse than s''

Code 10.7 tsp_GRASP_PR.py GRASP with path relinking implementation for the TSP

```
from random_generators import unif                              # Listing 12.1
from tsp_utilities import *                                      # Listing 12.2
from tsp_GRASP import tsp_GRASP                                  # Listing  7.4
from tsp_path_relinking import tsp_path_relinking                # Listing 10.6

######### GRASP with path relinking for the TSP
def tsp_GRASP_PR(d,                                           # Distance matrix
                 iterations,                         # Number of calls to GRASP
                 population_size,                     # Size of the population
                 alpha):                                      # GRASP parameter

    n = len(d[0])
    population = [[-1] * population_size for _ in range(population_size)]
    pop_size = iteration = 0
    lengths = [-1] * population_size
    while (pop_size < population_size and iteration < iterations):
        tour, tour_length = tsp_GRASP(d, alpha)
        iteration += 1
        succ = tsp_tour_to_succ(tour)
        different = True
        for i in range(pop_size - 1):
            if tsp_compare(population[i], succ) == 0:
                different = False
                break                          # The tour is already in population
        if different:
            population[pop_size] = succ[:]
            lengths[pop_size] = tour_length
            pop_size += 1
    if (iteration == iterations):#Unable to generate enough different solutions
        population_size = pop_size
    for it in range(iteration, iterations):
        tour, tour_length = tsp_GRASP(d, alpha)
        iteration += 1
        succ = tsp_tour_to_succ(tour)
        successors, length = tsp_path_relinking(d,
                     population[unif(0,population_size-1)],tour_length, succ)
        max_difference, replacing = -1, -1
        for i in range(population_size):
            if (length <= lengths[i]):
                difference = tsp_compare(population[i], successors)
                if difference == 0:
                    max_difference = 0
                    break
                if difference > max_difference and length < lengths[i]:
                    max_difference = difference
                    replacing = i
        if max_difference > 0:
            lengths[replacing] = length
            population[replacing] = successors[:]
            print('GRASP_PR population updated:', it, length)

    best = 0
    for i in range(1, population_size):
        if lengths[i] < lengths[best]:
            best = i
    return tsp_succ_to_tour(population[best]), lengths[best]
```

10.7 Fixed Set Search

The Fixed Set Search method [4] (FSS) also incorporates several mechanisms that are discussed in this book. First, a population of solutions is generated using a standard GRASP procedure. Then, this population is gradually improved by applying a GRASP procedure guided by a learning mechanism. The latter can be seen as a vocabulary building: one randomly selects a few solutions from the population and calculates the frequency of occurrence of the elements constituting these solutions. Then, another solution is randomly selected from the population. Among the elements constituting this solution, a fixed number are retained, determined by those which have the highest frequency of occurrence previously calculated. The randomized greedy construction is modified so that it produces a solution containing all the fixed elements.

In the case of the TSP, these elements form sub-paths. A step in the randomized construction adds either an edge connecting a city not in the selected sub-paths or all the edges of a fixed sub-path. The tour thus constructed is improved by a local search and enriches the population of solutions.

The FSS method has several parameters: a stopping criterion (e.g., a number of iterations without improvement of the best solution), the number of solutions selected to determine the fixed set, the number of elements of the fixed set (which can vary from one iteration to another), and the α parameter of the randomized construction.

Another way of looking at FSS is to see it as an LNS-type method (Section 6.4.1) with learning mechanisms: the acceptance method manages a population of solutions. The destruction method chooses a random solution from the population and relaxes the elements that do not appear frequently in a random sample of solutions from the population.

10.8 Particle Swarm

Particle swarms are a bit special because they were first designed for continuous optimization. The idea is to evolve a population of particles. Their position represents a solution to the problem expressed as a vector of real numbers. The particles interact with each other. Each has a velocity in addition to its position and is attracted or repelled by the other particles.

This type of method, proposed by Kennedy and Eberhart [5], simulates the behavior of animals living in swarms, such as birds, insects, or fish, which adopt a behavior that favors their survival, whether it be to feed, defend themselves against predators, or undertake a migration. Each individual in the swarm is influenced by those nearby and possibly by a leader.

Translated into optimization terms, each particle represents a solution to the problem whose quality is measured by a fitness function. A particle moves at

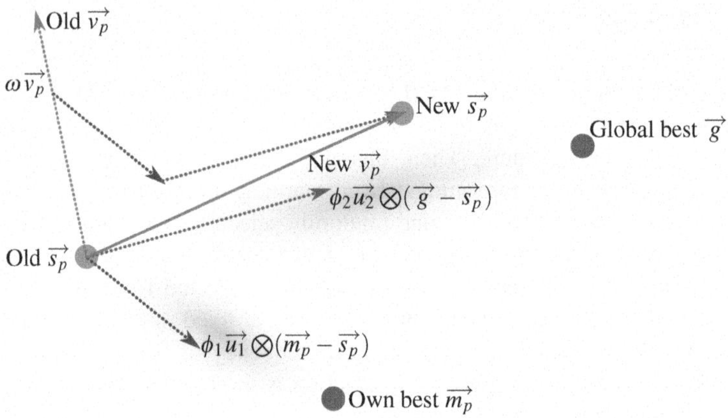

Fig. 10.14 Swarm particles velocity and position update

a certain speed in a given direction, but it is deflected by its environment: if there is another particle-solution of better quality in the vicinity, it is attracted in its direction. In that manner, each solution enumerated by the algorithm can be associated with the vertex of a graph. The edges of this graph correspond to particles that influence each other.

There are various variants of particle swarm methods, differing in the influence graph and the formulae used to calculate the deviations in particle velocity $\overrightarrow{v_p}$. In its most classic version, a particle p is influenced by only two solutions: the global best solution $\overrightarrow{g}$ found by the set of particles and the best solution $\overrightarrow{m_p}$ it has found itself. The new velocity of the particle is a vector. Each component is modified with weights randomly drawn between 0 and ϕ_1 in the direction of $\overrightarrow{m_p}$ and drawn between 0 and ϕ_2 in the direction of $\overrightarrow{g}$, where ϕ_1 and ϕ_2 are parameters of the method. In addition, a particle is given an inertia ω as a parameter. Figure 10.14 illustrates the update of the velocity and the position of a particle. Algorithm 10.4 provides a simple particle swarm framework.

Various modifications have been proposed to this basic version. For instance, instead of being influenced by the best solution it has found itself, a particle is influenced by the best solution found by particles in its neighborhood. It is then necessary to define according to which topology the latter is chosen. A common variant is to constrain the velocity of the particles to remain between two bounds, v_{min} and v_{max}. This adds two parameters to the framework. Another proposed modification is to apply a small random shift to some particles to simulate turbulence.

Algorithm 10.4: Swarm particle framework. $\otimes$ is component-wise multiplication

Input: Function $f : [\overrightarrow{x}_{min}, \overrightarrow{x}_{max}] \in \mathbb{R}^n \rightarrow \mathbb{R}$ to minimize, parameters $\mu, \omega, \phi_1, \phi_2, I_{max}$
Result: $\overrightarrow{g}$

$f^* = \infty$
for $p = 1 \ldots \mu$ **do**
    $\overrightarrow{v_p} \leftarrow \overrightarrow{unif}(\overrightarrow{x}_{min} - \overrightarrow{x}_{max}, \overrightarrow{x}_{max} - \overrightarrow{x}_{min})$ // Initial particle velocity
    $\overrightarrow{s_p} \leftarrow \overrightarrow{unif}(\overrightarrow{x}_{min}, \overrightarrow{x}_{max})$ // Initial particle position (solution)
    $\overrightarrow{m_p} \leftarrow \overrightarrow{s_p}$ // Best own position
    if $f^* > f(\overrightarrow{s_p})$ **then** Update gloal best
        $f^* \leftarrow f(\overrightarrow{s_p})$
        $\overrightarrow{g} \leftarrow \overrightarrow{s_p}$
for I_{max} *iterations* **do**
    for $p = 1 \ldots \mu$ **do**
        $\overrightarrow{u_1} \leftarrow \overrightarrow{unif}(\overrightarrow{0}, \overrightarrow{1})$
        $\overrightarrow{u_2} \leftarrow \overrightarrow{unif}(\overrightarrow{0}, \overrightarrow{1})$
        $\overrightarrow{v_p} \leftarrow \omega\overrightarrow{v_p} + \phi_1 \overrightarrow{u_1} \otimes (\overrightarrow{m_p} - \overrightarrow{s_p}) + \phi_2 \overrightarrow{u_2} \otimes (\overrightarrow{g} - \overrightarrow{s_p})$ // Update velocity
        $\overrightarrow{s_p} \leftarrow \overrightarrow{max}(\overrightarrow{min}(\overrightarrow{s_p} + \overrightarrow{v_p}, \overrightarrow{x}_{max}), \overrightarrow{x}_{min})$ // Update position
        if $f(\overrightarrow{m_p}) > f(\overrightarrow{s_p})$ **then** Update own best
            $\overrightarrow{m_p} \leftarrow \overrightarrow{s_p}$
            if $f^* > f(\overrightarrow{s_p})$ **then** Update gloal best
                $f^* \leftarrow f(\overrightarrow{s_p})$
                $\overrightarrow{g} \leftarrow \overrightarrow{s_p}$

10.8.1 Electromagnetic Method

In the electromagnetic method, a particle induces a force of attraction or repulsion on all the others. This force depends on the inverse of the square of the distance between the particles, like electrical forces. The direction of the force depends on the quality of the solutions. A particle is attracted by a solution that is better than itself and repelled by a worse solution.

10.8.2 Bestiary

In previous sections, we have only mentioned the basic algorithms, inspired by the behavior of social animals, and a variant, inspired by a process of physics. Different authors have proposed many metaheuristics whose framework is similar to that of Algorithm 10.4.

What distinguishes them is essentially the way of initializing the speed and the position of the particles (lines 3 and 4) as well as the "magic formulas" for their updates (lines 13 and 14).

These various magical formulas are inspired by the behavior of various animal species or in the processes of physics. To name just a few, there are amoeba, bacteria, bat, bee, butterfly, cockroaches, cuckoo, electromagnetism, firefly, and mosquito. There are various variants of these frameworks, obtained by hybridizing them with the key components of the metaheuristics discussed in this book. There are hundreds of proposals in the literature suggesting "new" metaheuristics inspired by various metaphors, sometimes even referring to the behavior of mythic creatures!

Very schematically, it is a matter of applying the intensification and diversification principles: elimination of certain solutions from the population, concentration toward the best discovered solutions, random walk, etc.

A number of these frameworks have been proposed in the context of continuous optimization. To adapt these methods to discrete optimization, one can implement a coding scheme, for example, the random keys seen in Section 10.5. Another solution is to consider the notion of neighborhood and path relinking. The reader who is a friend of animals and other creatures may consult [7] for a bestiary overview.

Rather than trying to devise a new heuristic based on an exotic metaphor using obscure terminology, we encourage the reader to use a standardized description, following the basic principles presented in this book. Indeed, during the last quarter century, there have been few truly innovative new concepts. It is a matter of adopting a more scientific posture, of justifying the choices of problem modeling, of establishing test protocols, etc., even if the development of a theory of metaheuristics still seems very far away and the heuristic solution of real-world optimization problems remains the only option.

Problems

10.1 Genetic Algorithm for a One-Dimensional Function

We need to optimize a function f of an integer variable x, $0 \leqslant x < 2^n$. In the context of a genetic algorithm with a standard crossover operator, how to encode x in the form of a binary vector?

10.2 Inversion Sequence

A permutation p of elements from 1 to n can be represented by an *inversion sequence* s, where s_i counts the number of elements of $p_1, \dots, p_k = i$ that are greater than i. For example, the permutation $p = (2, 4, 6, 1, 5, 3)$ has the inversion sequence $s = (3, 0, 3, 0, 1, 0)$: there are three elements greater than 1 before 1, 0 elements greater than 2 before 2, etc. To which permutations do the inversion sequences $(4, 2, 3, 0, 1, 0)$ and $(0, 0, 3, 1, 2, 0)$ correspond? Provide necessary and sufficient conditions for a vector s to be an inversion sequence corresponding to a permutation. Can the standard 1-point, 2-point, and uniform crossover operators be applied to inversion sequences? How can inversion sequences be used in the context of scatter search?

10.3 Rank Based Selection
What is the probability of the function `rank_based_selection(m)`, given in Algorithm 10.1, to return a given value v?

10.4 Tuning a Genetic Algorithm
Adjust the population size and mutation rate of the procedure `tsp_GA` given by Code 10.5, if it generates a total of $5n$ children.

10.5 Scatter Search for the Knapsack Problem
Consider the knapsack instance 10.1 of Section 10.4. Perform the first iteration of a scatter search for this instance: generate the new population, repair/improve the solutions, update the reference set consisting of five solutions with three elites.

References

1. Glover, F.: Heuristics for integer programming using surrogate constraints. Decision Sciences **8**(1), 156–166 (1977). https://doi.org/10.1111/j.1540-5915.1977.tb01074.x
2. Glover, F.: Tabu search and adaptive memory programming — advances, applications and challenges. In: Barr, R.S., Helgason, R.V., Kennington, J.L. (eds.) Interfaces in Computer Science and Operations Research: Advances in Metaheuristics, Optimization, and Stochastic Modeling Technologies. pp. 1–75. Springer, Boston (1997). https://doi.org/10.1007/978-1-4615-4102-8_1
3. Holland, J.: Adaptation in Natural and Artificial Systems. The University of Michigan Press, Ann Harbor (1975)
4. Jovanovic, R., Tuba, M., Voß, S.: Fixed set search applied to the Traveling Salesman Problem. In: Blum, C., Gambini Santos, H., Pinacho-Davidson, P., Godoy del Campo, J. (eds.) Hybrid Metaheuristics. Lecture Notes in Computer Science, vol. 11299, pp. 63–77. Springer, Cham (2019). https://doi.org/10.1007/978-3-030-05983-5_5
5. Kennedy, J., Eberhart, R.C.: Particle swarm optimization. In: IEEE International Conference on Neural Networks, vol. IV, pp. 1942–1948. IEEE Service Center, Perth, Piscataway, NJ (1995). https://doi.org/10.1109/ICNN.1995.488968
6. Laguna, M., Martí, R.: GRASP and path relinking for 2-layer straight line crossing minimization. INFORMS J. Comput. **11**(1), 44–52 (1999). https://doi.org/10.1287/ijoc.11.1.44
7. Lones, M.: Mitigating metaphors: A comprehensible guide to recent nature-inspired algorithms. SN Comput. Sci. **1**(49), (2020). https://doi.org/10.1007/s42979-019-0050-8
8. Moscato, P.: Memetic algorithms: A short introduction. In: Corne, D., Glover, F., Dorigo, M. (eds.) New Ideas in Optimisation, pp. 219–235. McGraw-Hill, London (1999)
9. Rechenberg, I.: Evolutionsstrategie: Optimierung technischer Systeme nach Prinzipien der biologischen Evolution. Frommann-Holzboog, Stuttgart (1973). https://doi.org/10.1002/fedr.19750860506
10. Resende, M.G.C., Riberio, C.C.: Optimization by GRASP: Greedy Randomized Adaptive Search Procedures. Springer, New York, (2016). https://doi.org/10.1007/978-1-4939-6530-4
11. Sörensen, K., Seveaux, M.: MA|PM: Memetic algorithms with population management. Comput. Oper. Res. **33**, 1214–1225 (2006). https://doi.org/10.1016/j.cor.2004.09.011

Chapter 11
Heuristics Design

This chapter gives some tips for developing heuristics. It goes back to the modeling of the problem and gives an example of decomposing the problem into a chain of sub-problems that are easier to treat. It then proposes an approach to designing a specific heuristic. Finally, techniques for parameter tuning and comparison of algorithms are discussed.

11.1 Problem Modeling

Now that we have reviewed the key ingredients of heuristics, let us try to propose an approach to design one. The first thing to determine is whether a heuristic approach is absolutely necessary. Indeed, the "no free lunch" theorem [5] informs us that no optimization heuristics outperforms all others! This result follows from the fact that in the infinite variety of instance data, most of them have no exploitable structure. Any heuristic, no matter how sophisticated, selects therefore an unfortunate choice for a given data set. Among the infinite number of imaginable heuristics, there is at least one that does not include this inappropriate choice.

If one has to solve a concrete optimization problem, it is therefore necessary to "cheat." Examples are the set bipartition and the knapsack problems discussed in Sect. 1.2.3.5. If we know that the data do not contain large values, it is useless to design a heuristic because we can solve this type of problem exactly by dynamic programming.

The way in which the problem is modeled is crucial to its successful resolution. Especially when dealing with a concrete logistics problem, it can be time-consuming and tedious to capture all the wishes of a manager who is not used to the narrow view of a optimization engineer, who thinks in terms of variables, objective function, and constraints.

É. D. Taillard, *Design of Heuristic Algorithms for Hard Optimization*, Graduate Texts in Operations Research, https://doi.org/10.1007/978-3-031-13714-3_11

To do this, one must first identify the variables of the problem. One must determine what can be modified and what is intangible and part of the data. Then, the constraints must be discussed, specifying those that are hard and must be respected for a solution to be operational. Often, constraints presented as indispensable correspond rather to good practices, from which it is possible to depart from time to time. These soft constraints are generally integrated into an objective with a penalty factor. As we have seen with the Lagrangian relaxation technique, hard constraints can also be introduced into an objective, but probably with a higher penalty weighting. Finally, in practice, there are several objectives to be optimized. However, a manager may not be very happy if provided with a huge set of Pareto optimal solutions (see Problem 11.1) and has to examine all of them before choosing one. On the other hand, it will be easier to prioritize the objectives. It remains to be seen whether these objectives should be treated in a hierarchical manner (the optimum of the highest priority objective is sought before optimizing the next one) or by scalarization (all the objectives are aggregated into one, with weights in relation to their priority).

Once the problem has been properly identified, the designer of a heuristic algorithm must choose a model that is appropriate for the solution method. The following section illustrates two remarkably similar models of the same problem that can lead to the design of very different algorithms.

11.1.1 Model Choice

To reconstruct an unknown genetic sequence, a DNA microarray chip, able to react to all k-nucleotide sequences, is exposed to the gene to be discovered. Once revealed, this chip allows knowing all subsequences of k-nucleotides present in the gene to be analysed.

The data can be modeled using de Bruijn graphs. These graphs represent the superposition of symbol chains.

A first model associates an arc with each k-nucleotide detected. So, the 3-nucleotide AAC is represented by an arc connecting the vertices $AA \rightarrow AC$, due to the middle A superposition. If m is the number of k nucleotides detected, we have a graph with m edges. The reconstruction problem is to find an Eulerian path (passing through all the arcs) in this graph. This problem is easy, it can be solved in linear time.

The other model associates a vertex with each k-nucleotide detected. An arc connects two vertices if the associated k-nucleotides have a common subsequence of $k - 1$ nucleotides. For instance, if the 3-nucleotides AAC, ACA and ACG are detected, then both arcs $AAC \rightarrow ACA$ and $AAC \rightarrow ACG$ are present in the graph due to the common AC superposition. The graph is a directed version of the line graph of the previous representation. The reconstruction problem is to discover a Hamiltonian path (passing once through all the vertices). This second modeling thus requires the resolution of an NP-complete problem.

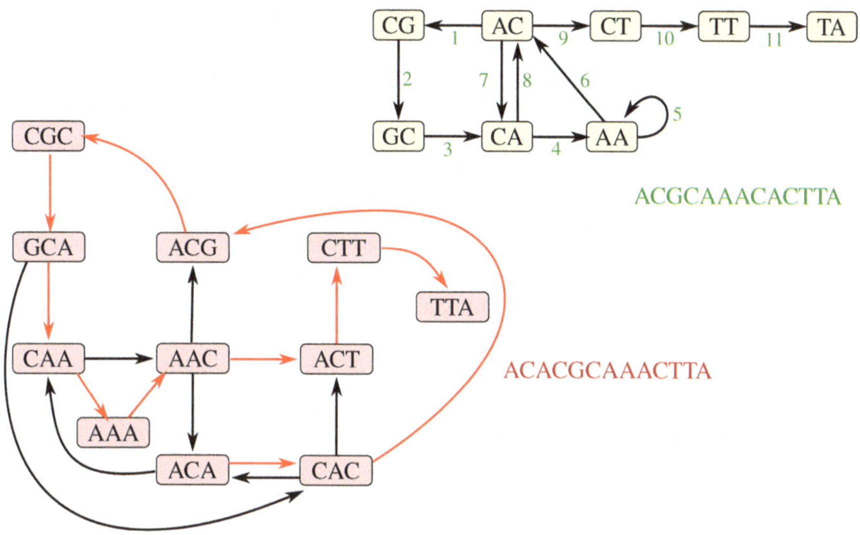

Fig. 11.1 Two de Bruijn graph models for the reconstruction of a genetic sequence. Top right, a graph in which an arc connecting two $k-1$-nucleotides represents a detected k-nucleotide (with $k = 3$). With this model, we have to find a Euclidean path in the graph. The numbering of the arcs corresponds to one of the shortest possible sequences. Bottom left is a graph in which the vertices represent the detected k-nucleotides. An arc represents a $k+1$-nucleotide that could be in the sequence. With this model, we have to discover a Hamiltonian path. The colored arcs provide the path of the other shortest sequence

That said, this second model is not necessarily to be discarded in practice. Indeed, a concrete sequencing problem is likely to possess peculiarities and constraints that might be more difficult to deal with using the first model. For example, a genetic sequence may include repeated subsequences, a more or less reliable quantification of the number of times a k-nucleotide appears, etc.

Figure 11.1 shows the graphs that can be constructed with these two models for a gene that has activated 11 subsequences of 3-nucleotides. In this example, it is not possible to unambiguously reconstruct the gene.

The choice of a model is often sensitive and depends on the techniques to be implemented. During the design of a heuristic, it is frequent to realize that another model is more appropriate. In any case, it is noteworthy to keep in mind that good solutions are located at the boundary between feasible and unfeasible ones. To highlight the point, the 2-opt neighborhood for the TSP is restricted to examining feasible solutions. Despite constituting the primary operations of the Lin-Kernighan neighborhood, it is much less efficient. The success of the latter is undoubtedly due to the fact that one examines reference structures that are not tours. In a way, feasible solutions are approached from outside the domain of definition.

When a few constraints are numerical, the possibility of implementing the Lagrangian relaxation technique discussed in Sect. 2.8 should be studied. By adjusting the value of the penalties associated with the violation of the relaxed constraints, the heuristic focuses its search in the vicinity of the boundary of feasible solutions.

Conversely, artificial constraints can also be added to implement the diversification technique outlined in Sect. 9.2. For example, the quality of the solutions generated by Algorithms 2.7 and 2.8 for clustering can be significantly improved by adding a soft constraint imposing that the groups must contain the same number of elements. This constraint is relaxed and introduced in the fitness function with a penalty parameter that decreases during the iterations, just like the temperature of a simulated annealing. Thus, by performing a few more iterations than the original method, the solution is perturbed by making the largest groups smaller and the very small groups larger. At the end of the algorithm, the penalties being very low, we return to the initial objective, but the centers have managed to better relocate. The local optimum thus obtained is considerably better than that of the original method which does not move significantly enough from the initial random solution.

11.1.2 Decomposition into a Series of Sub-problems

Another step in modeling is to assess whether it is possible to apply the *divide and conquer* principle. Rather than trying to design a complex problem model with multiple interrelationships, one can try breaking it down into a series of more easily addressed sub-problems.

An example is the vehicle routing problem and its extension to several warehouses that need to be positioned. Rather than solving the positioning of the warehouses and the construction of the routes simultaneously, one can initially only deal with the customers constituting natural groups. The creation of the routes can be done in a second step and then finally the positioning of the warehouses. Figure 11.2 illustrates the process of this decomposition into a succession of sub-problems. With this approach, it was possible to handle examples with millions of elements with a reasonable computational effort. At the time these results were obtained, the instances in the literature were 1000 or 10,000 times smaller.

11.2 Algorithmic Construction

Once the problem has been modeled and a fitness function has been chosen, the construction of an algorithm can start. The first step is to construct a solution. Chapter 4 suggests diverse ideas to realize this step. Beforehand, if the instance size is significant, it is necessary to consider another problem partitioning, by the data.

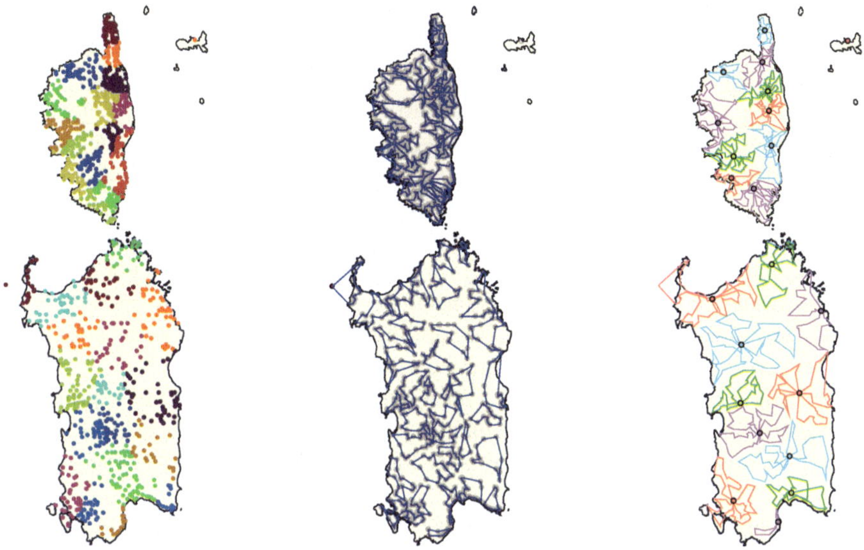

(a) (b) (c)

Fig. 11.2 Decomposition of the location-routing problem into a sequence of sub-problems. (**a**) Generation of customer clusters with a rapid clustering method. (**b**) Generation of routes for each customer cluster. (**c**) Positioning of warehouses and re-optimization of routes

11.2.1 Data Slicing

A partition of the problem items should be considered if the volume of data is very large or if a procedure of high algorithmic complexity is to be implemented. For the location-routing problem discussed in the previous section, a proximity graph presented in Sect. 6.3.1 was used. The size of the clusters created was chosen according to the problem data, so that their volume was close to a (small) multiple of the volume of the vehicles. The routing heuristic has also been designed in anticipation of optimizing a few tours at a time. It may therefore be appropriate to slice the data even for relatively small instances.

11.2.2 Local Search Design

Purely constructive heuristic algorithms rarely produce solutions of acceptable quality for difficult optimization problems. Even when combined with learning methods such as pheromone trails in artificial ant colonies or population management with genetic algorithms, convergence can be slow and the resulting solutions of insufficient quality. However, some software libraries can automatically generate a genetic algorithm with only the decoding of a solution from a binary vector and the calculation of the fitness function. In certain situations, the coding phase of the algorithm can be significantly reduced.

Generally, it is essential to move on to the subsequent phase and devise a neighborhood for the problem to be solved. The success of a heuristic algorithm frequently depends on the design of the envisaged neighborhood(s). Metaheuristics are sometimes described as processes guiding a local search. This explains the relative length of Chap. 5 and the richness of the various techniques allowing their limitation or extension. However, we are consciously aware that not everyone possesses the genius of Lin and Kernighan to design such an efficient ejection chain for the traveling salesman problem. Perhaps the latter is an exception, as it does not need to rely on other concepts to achieve excellent quality solutions.

As it is usually not possible to find a neighborhood with all the appropriate characteristics (connectivity, small diameter, fast evaluation, etc.), a local search often uses several different neighborhoods. Each of them corrects a weakness of the others.

This implies thinking about strategies for their use: one has to decide whether the local search should evaluate all these neighborhoods at each iteration or whether one should alternate phases of intensification, using one neighborhood, and diversification of the search, using other neighborhoods.

11.3 Heuristics Tuning

Theoretically, a programmer who is not an expert on the problem to be solved should be able to design a heuristic based on the key principles of metaheuristics discussed in the previous chapters. In practice, during the algorithm development, the programmer is going to gain some experience on how to achieve good solutions for the specific type of data to be processed.

Indeed, the most time-consuming work in the design of a heuristic algorithm consists in trying to understand why the constructive method goes wrong and produces outliers, why the "genial" neighborhood does not give the predicted results or why the "infallible" learning method fails…

11.3.1 Instance Selection

The design of a heuristic algorithm begins with the selection of the problem instances that it should be able to successfully tackle. Indeed, the "no free lunch" theorem tells us it is an illusion to try to design a universally successful method.

When addressing an industrial problem, once we have managed to translate the customer's wishes into a model that seems reasonable, we still need to obtain concrete numerical data. This sometimes problematic phase can require a disproportionate investment, especially if the model developed is too complex, with poor decomposition and not sufficiently focused on the core of the problem. In practice, it frequently occurs that constraints described as essential are not imperative but result from a fear of altering too drastically the current operating mode and habits. Conversely, the first feedback from solutions provided by a heuristic algorithm may also reveal constraints that have not been explicitly stated. In this case, it is necessary to go back to the modeling stage and repeat an iteration...

For an academic problem, there are usually libraries with many numerical data. In this case, a selection of the instances must be considered so as not to invest an infinite amount of time tuning the algorithm. To evaluate the proper implementation of the developed heuristic, a first selection should consider moderate size instances for which an optimal or a very good solution is identified.

The instance selection should also be able to highlight the pathological cases for the developed heuristic. This choice must also be governed by the interest of these examples for practical cases. One example is the case of the satisfiability problems with 3 literals per clause (3SAT). If the number of randomly generated clauses is significant compared to the number of variables, then the instances are easy: the probability that there is a feasible assignment of the variables tends very quickly to 0. Conversely, if the number of clauses is limited compared to the number of variables, then the examples are equally easy: there is a probability tending very quickly toward 1 that there is a feasible assignment. It was determined that the transition between intractable and simple instances occurs when the number of clauses is 4.24 times higher than the number of variables. This result is interesting in itself, and while an efficient heuristic is developed for this type of instances, it does not guarantee it will be efficient for practical applications.

Finally, results for problem instances with very diverse characteristics should be separately reported. Indeed, multiplying the number of favorable (e.g., very small) or unfavorable instances would lead to biased results.

11.3.2 Graphical Representation

When possible, a graphical representation of the solutions helps to perceive how a heuristic works and to correct its flaws. Indeed, it is frequent to imagine bad causes explaining poor results. By visualizing the output of the method as something other than a series of numbers, it is sometimes very simple to explain this poor performance.

For some problems, a graphical representation is natural, as for the Euclidean TSP. This certainly explains the excellent efficiency of the heuristic and exact methods that have been developed for this problem.

11.3.3 Parameter and Option Tuning

The most time-consuming step in the design of a heuristic is the selection of its ingredients and parameter tuning. When developing a heuristic, one desires it to provide results of the highest possible quality on as wide an instance range as possible.

Initially, the programmer can proceed intuitively to find parameter values that fulfil this purpose. Typically, a small instance is chosen and the heuristic is executed by varying parameter values or changing options. The most promising evolutions are favored. Put differently, the tuning consists in applying a heuristic to another problem whose variables are the parameter values and whose objective function is the result of the run of the program implementing the algorithm under development.

In principle, the search space for tuning variables is considerably smaller than that of the instance to be solved. If not, the question arises as to the relevance of a heuristic whose design could be more complicated than the problem to be solved.

Several illustrations in this book provide the results of extensive experiments on the influence of the value of one or two parameters for some heuristics. For example, Fig. 9.3 shows that for a TSP instance with $n = 127$ cities, the appropriate combinations of the parameters d_{min} and Δ seem to be such that $\Delta + 2d_{min} = n$, provided that one performs $10n$ iterations of the taboo search.

These results are not intended to provide definitive values. They are presented so that the reader can get an idea of the appropriate values, but they are not necessarily generalizable. The production of such a figure requires a disproportionate effort (more than 10,000 executions of the heuristic and then production of the diagram) compared to the information that can be obtained. However, it does allow us to observe significant random fluctuations in the results obtained.

If the heuristic has more than half a dozen parameters and options, a rough intuitive tuning is likely to be biased:

- Given the effort involved, few alternatives are tested.
- The instance set is limited.
- The heuristic only works well on a limited instance set.
- Outstanding or bad results focus attention.
- Results are neither reproducible nor statistically supported.

It is therefore recommended to use automated methods to calibrate the parameters, providing these methods with a sufficiently representative instance set. They have the advantage of not being subjective, focusing on a set that is very favorable or leaving out a very unfavorable instance. As a parameter adjustment software, we can quote, among others, *iRace* proposed by [2].

11.3.4 Measure Criterion

The design of heuristics is habitually a multi-objective process. Indeed, the vast majority of the framework algorithms discussed in Part III of this book include a parameter that directly influences the number of repetitions of a general loop. Consequently, one can choose the computational effort to solve a problem quite freely. Furthermore, the quality of the solution produced depends directly on this effort. An extreme case is given by simulated annealing, which almost certainly produces the best possible solution, provided that one accepts an infinite effort!

A compromise must therefore be achieved between the computational time and the quality of the solutions produced.

11.3.4.1 Success Rate

A first measure of the quality of a heuristic is its success rate in producing target solutions. These may be the optimum, if known, or solutions of a given quality. If the value of the optimum is unknown, a bound can be derived using a relaxation, from which a certain deviation can be accepted.

The simplest case for comparing success rates occurs when the heuristics have no parameters, or when the parameters have been fixed. In this case, we want to answer the question: does heuristic $\mathcal{A}$ find more target solutions than heuristic $\mathcal{B}$? The answer to this question is univocal: we run $\mathcal{A}$ and $\mathcal{B}$ on the same set of instances and we count the number of respective successes. Obviously, for this to make sense, the instances must be chosen prior to the experiment, and not according to the results obtained by one or the other method.

As in any experiment of this type, a subsidiary question must be answered: is the observed difference in success rates significant? Indeed, if the heuristics include a random component, or if the instance set is randomly selected, the difference could be due to chance and not to a distinct solving performance between the heuristics.

In this case, a statistical test can be carried out, with the null hypothesis that both methods have exactly the same probability p of success [4]. To conduct such a test, the independence of the experiments should be guaranteed. Under these conditions, relatively few numerical experiments can reveal a significant difference.

Table 11.1 provides the values for which it can be stated with 99% confidence that one proportion is significantly higher than another. This table can be used as follows: suppose we want to compare the $\mathcal{A}$ and $\mathcal{B}$ methods on instances drawn at random (e.g., TSPs with 100 cities uniformly generated in the unit square). Suppose that the $\mathcal{B}$ method was able to find a solution of given quality only twice on $n_b = 5$ runs. Suppose that, out of $n_a = 10$ runs of the method $\mathcal{A}$, 9 were able to achieve such quality. In the corresponding row and column of Table 11.1, we find the couple $(10, 2)$. In other words, a proportion of at least $10/10$ should have been reached to conclude that the $\mathcal{A}$ method is superior to the $\mathcal{B}$ method.

Table 11.1 Pairs (a, b) for which a success rate $\geq a/n_a$ is significantly higher than a rate $\leq b/n_b$, for a confidence level of 99%

		n_a													
		2	3	4	5	6	7	8	9	10	11	12	13	14	15
n_b	2						(7,0)	(8,0)	(9,0)	(10,0)	(11,0)	(12,0)	(12,0)	(13,0)	(14,0)
	3			(4,0)	(5,0)	(6,0)	(7,0)	(7,0)	(8,0)	(9,0)	(9,0)	(10,0)	(11,0)	(11,0)	(12,0)
												(12,1)	(13,1)	(14,1)	(15,1)
	4		(3,0)	(4,0)	(5,0)	(5,0)	(6,0)	(6,0)	(7,0)	(8,0)	(8,0)	(9,0)	(9,0)	(10,0)	(11,0)
						(6,1)	(7,1)	(8,1)	(9,1)	(10,1)	(11,1)	(11,1)	(12,1)	(13,1)	(14,1)
	5		(3,0)	(4,0)	(4,0)	(5,0)	(5,0)	(6,0)	(6,0)	(7,0)	(7,0)	(8,0)	(8,0)	(9,0)	(9,0)
					(5,1)	(6,1)	(7,1)	(7,1)	(8,1)	(9,1)	(10,1)	(10,1)	(11,1)	(12,1)	(12,1)
									(9,2)	(10,2)	(11,2)	(12,2)	(13,2)	(14,2)	(14,2)
	6		(3,0)	(3,0)	(4,0)	(4,0)	(5,0)	(5,0)	(6,0)	(6,0)	(7,0)	(7,0)	(8,0)	(8,0)	(9,0)
				(4,1)	(5,1)	(6,1)	(6,1)	(7,1)	(8,1)	(8,1)	(9,1)	(9,1)	(10,1)	(11,1)	(11,1)
						(6,2)	(7,2)	(8,2)	(9,2)	(10,2)	(10,2)	(11,2)	(12,2)	(13,2)	(13,2)
											(11,3)	(12,3)	(13,3)	(14,3)	(15,3)
	7	(2,0)	(3,0)	(3,0)	(4,0)	(4,0)	(5,0)	(5,0)	(5,0)	(6,0)	(6,0)	(7,0)	(7,0)	(7,0)	(8,0)
				(4,1)	(5,1)	(5,1)	(6,1)	(6,1)	(7,1)	(8,1)	(8,1)	(9,1)	(9,1)	(10,1)	(10,1)
					(5,2)	(6,2)	(7,2)	(8,2)	(8,2)	(9,2)	(10,2)	(10,2)	(11,2)	(12,2)	(12,2)
								(8,3)	(9,3)	(10,3)	(11,3)	(12,3)	(12,3)	(13,3)	(14,3)
													(13,4)	(14,4)	(15,4)

8	(2,0)	(3,0)	(3,0)	(4,0)	(4,0)	(4,0)	(5,0)	(5,0)	(5,0)	(6,0)	(6,0)	(6,0)	(7,0)	(7,0)
		(3,1)	(4,1)	(4,1)	(5,1)	(6,1)	(6,1)	(7,1)	(7,1)	(8,1)	(8,1)	(9,1)	(9,1)	(10,1)
			(4,2)	(5,2)	(6,2)	(6,2)	(7,2)	(8,2)	(8,2)	(9,2)	(10,2)	(10,2)	(11,2)	(12,2)
					(6,3)	(7,3)	(8,3)	(9,3)	(9,3)	(10,3)	(11,3)	(12,3)	(12,3)	(13,3)
								(9,4)	(10,4)	(11,4)	(12,4)	(13,4)	(14,4)	(14,4)
														(15,5)
9	(2,0)	(3,0)	(3,0)	(3,0)	(4,0)	(4,0)	(4,0)	(5,0)	(5,0)	(5,0)	(6,0)	(6,0)	(6,0)	(7,0)
		(3,1)	(4,1)	(4,1)	(5,1)	(5,1)	(6,1)	(6,1)	(7,1)	(7,1)	(8,1)	(8,1)	(9,1)	(9,1)
			(4,2)	(5,2)	(6,2)	(6,2)	(7,2)	(7,2)	(8,2)	(9,2)	(9,2)	(10,2)	(10,2)	(11,2)
				(5,3)	(6,3)	(7,3)	(8,3)	(8,3)	(9,3)	(10,3)	(10,3)	(11,3)	(12,3)	(12,3)
						(7,4)	(8,4)	(9,4)	(10,4)	(11,4)	(11,4)	(12,4)	(13,4)	(14,4)
									(10,5)	(11,5)	(12,5)	(13,5)	(14,5)	(15,5)
10	(2,0)	(3,0)	(3,0)	(3,0)	(4,0)	(4,0)	(4,0)	(4,0)	(5,0)	(5,0)	(5,0)	(6,0)	(6,0)	(6,0)
		(3,1)	(4,1)	(4,1)	(5,1)	(5,1)	(5,1)	(6,1)	(6,1)	(7,1)	(7,1)	(8,1)	(8,1)	(9,1)
			(4,2)	(5,2)	(5,2)	(6,2)	(6,2)	(7,2)	(8,2)	(8,2)	(9,2)	(9,2)	(10,2)	(10,2)
				(5,3)	(6,3)	(7,3)	(7,3)	(8,3)	(9,3)	(9,3)	(10,3)	(10,3)	(11,3)	(12,3)
					(6,4)	(7,4)	(8,4)	(9,4)	(9,4)	(10,4)	(11,4)	(12,4)	(12,4)	(13,4)
							(8,5)	(9,5)	(10,5)	(11,5)	(12,5)	(13,5)	(13,5)	(14,5)
											(12,6)	(13,6)	(14,6)	(15,6)
11	(2,0)	(3,0)	(3,0)	(3,0)	(3,0)	(4,0)	(4,0)	(4,0)	(5,0)	(5,0)	(5,0)	(5,0)	(6,0)	(6,0)
		(3,1)	(4,1)	(4,1)	(4,1)	(5,1)	(5,1)	(6,1)	(6,1)	(6,1)	(7,1)	(7,1)	(8,1)	(8,1)
		(3,2)	(4,2)	(5,2)	(5,2)	(6,2)	(6,2)	(7,2)	(7,2)	(8,2)	(8,2)	(9,2)	(9,2)	(10,2)
			(4,3)	(5,3)	(6,3)	(6,3)	(7,3)	(8,3)	(8,3)	(9,3)	(9,3)	(10,3)	(11,3)	(11,3)
				(5,4)	(6,4)	(7,4)	(8,4)	(8,4)	(9,4)	(10,4)	(10,4)	(11,4)	(12,4)	(12,4)
						(7,5)	(8,5)	(9,5)	(10,5)	(10,5)	(11,5)	(12,5)	(13,5)	(13,5)
								(9,6)	(10,6)	(11,6)	(12,6)	(13,6)	(14,6)	(14,6)
												(13,7)	(14,7)	(15,7)

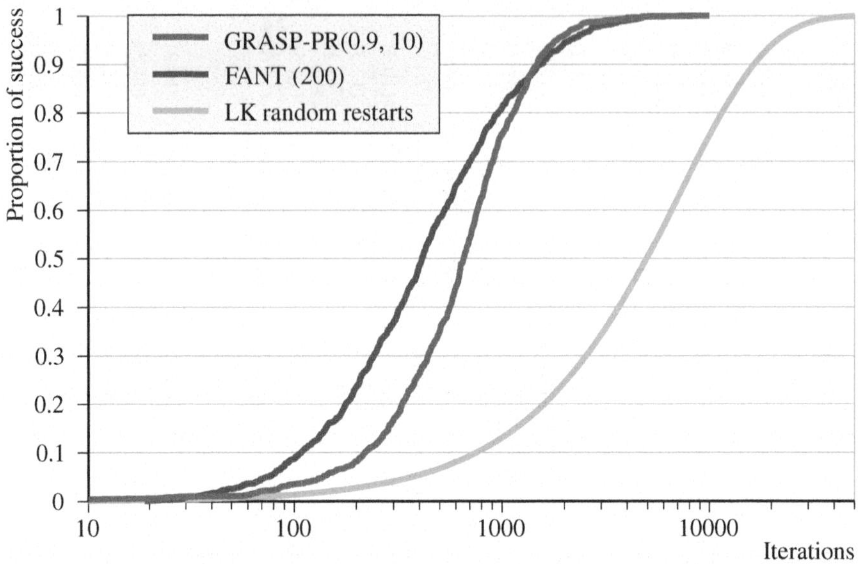

Fig. 11.3 Success rate for the optimal resolution of the TSP instance tsp225 by three iterative methods. The learning processes of the FANT (Code 8.3) and GRASP-PR (Code 10.7) methods demonstrate a reasonable efficiency. Indeed, independent runs of Code 12.3, starting from randomly generated solutions, present a much lower success rate

Put differently, there is a probability p of success such that in more than 1% of cases, we can observe nine successes out of ten or two successes out of five. This result is counterintuitive, as the observed success rates vary by a factor larger than two.

The situation becomes more complex if the success rate depends on the computational effort involved to reach a target solution. One possibility is to plot the results as a proportion of success versus effort (time-to-target plot or TTT-plot). Figure 11.3 illustrates this for three different heuristics with a small TSP instance. For this figure, the reference for an iteration represents a call to Code 12.3. The generation time of the starting solution has been neglected here. The last is generated either in a purely random way, or with an artificial pheromone matrix, or with a randomized greedy algorithm. We also ignore that the local search may take more or less time to complete, depending on the starting solution.

The success rate curve for the method repeating local searches from randomly generated solutions was obtained by estimating the probability of a successful run. This estimation required 100,000 executions of the method, with only 14 achieving the target.

The success rate of x executions was therefore estimated to be $1-(1-0.00014)^x$. However, this mode of representation is questionable, as the target value and the instance chosen can greatly influence the results. Moreover, the compared methods should require an approximately identical computational effort for each iteration.

11.3.4.2 Computational Time Measure

Whenever possible, one should always favor an absolute measure of the computational effort, for example, by counting a number of iterations. Obviously, one must specify what can influence the computational effort, typically the size of the data to be processed. The algorithmic complexity of an iteration should therefore be indicated.

This complexity is sometimes not clearly identifiable, or its theoretical expression has nothing to do with practical observations. The simplex algorithm for linear programming has already been mentioned, which can theoretically perform an exponential number of pivots, but in practice stops after an almost linear number of steps.

To compare the speed of heuristics, one is sometimes forced to use a relative measure, the computational time. For the same algorithm using the same data structures with the same algorithmic complexity, the computation time depends on many factors, among which:

- The programming language used for its implementation
- The hardware (processor, memory, etc.)
- The programming style
- The interpreter or compiler
- The interpretation or compilation options
- The operating system
- Running the system in energy-saving mode
- Other independent processes running in parallel
- The BIOS configuration

Achieving reliable computing times can represent a challenge. For example, the motherboards of personal computers are often configured from the factory to run in "turbo" mode. In practice, when the processor is not heavily used, the clock frequency drops, which reduces energy consumption. When starting intensive computations, the first iterations can therefore take much longer than the following ones, although they perform the same number of operations. The maximum clock frequency may depend on the fact that a laptop works on battery or with an external power supply.

Thus, a factor of 2 can indeed be observed for the execution of a procedure on two machines with the same hardware (or even on the same machine). The factor can rise to more than 100 if we compare two similar implementations but not using the same programming language.

To obtain meaningful results, it is frequently necessary to repeat runs for the same set of parameters if the measured times are less than 1 second. In all cases, it should be kept in mind that the computational time remains a relative measure. What is important is the evolution of time according to the characteristics of the problems being solved. One essential characteristic is the data size.

In Fig. 11.4, we have plotted the running time of some codes proposed in this book as a function of the number of cities in the problem. This figure uses two

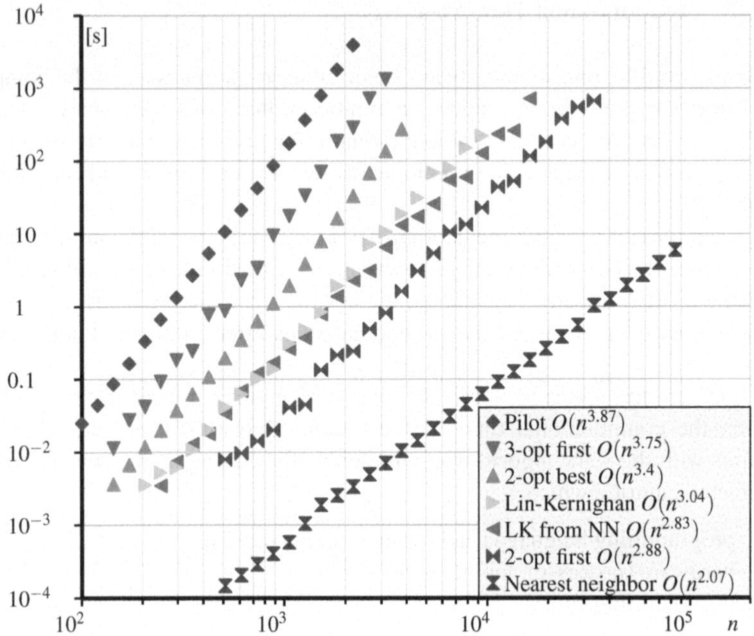

Fig. 11.4 Evolution of the computational time as a function of the number of TSP cities for some codes presented in this book

logarithmic scales. In this way, a polynomial growth of the computational time is represented, asymptotically, by a straight line. The slope of this line indicates the degree of the polynomial. It can be noted that all methods behave polynomially.

The reader who wishes to reproduce this figure should be able to do so without too much difficulty using the codes provided. The degree of the polynomials observed is likely to be close to that presented, but the time scale may be very different depending on the programming language and the configuration of the computer used.

11.3.4.3 Solution Quality Measure

Comparing the quality of solutions produced by fully determined heuristics (with no free parameters) is relatively simple. Each heuristic is run on one or more instance sets with similar characteristics, and the value of the objective functions produced is recorded. The most standard measure is certainly the average. However, to know if the average of the values produced by two heuristics is significantly different requires a statistical test. If we can reasonably assume that the values produced are normally distributed, there are standard tests that are relatively simple to implement, such as student's t test. If there are more than two methods to be compared, a more elaborate test, typically an analysis of variance (ANOVA), must be conducted.

If the values produced do not follow a normal distribution, particular caution should be observed, as a single measure can significantly alter an average. In this case, it is more reliable to use another measure, for example, the median, with Friedman's analysis of variance. With this test, a rank is assigned for each run of a method and the null hypothesis states that all samples are from a population with the same median.

Bootstrapping is a very general statistical technique that is fairly simple to implement and is particularly suitable for comparing methods for which it is not reasonable to obtain a large number of runs. The estimation of a quantity such as the mean, median, or their confidence interval is done by drawing a large number of samples from the small number of observations made. The quantity to be estimated for each of these samples is calculated, and the mean of the samples provides an estimator of this quantity. To obtain a confidence interval, it is sufficient to identify the quantiles of the resampling distribution.

When the heuristics are not completely determined, for example, if one wishes to provide a more complete picture of the evolution of the results as a function of the computational effort, the statistical tests mentioned above must be repeated for each computational effort. There are convenient tools to perform this automatically and provide diagrams with confidence intervals.

Figure 11.5 illustrates the evolution of the objective function for the same methods as in Fig. 11.3. As all three methods were run on the same machine, it is possible to compare them on the basis of computational time. This figure gives a double scale for the abscissa—Computational time, number of calls to the descent to a local optimum. For this diagram, the reference scale is time. The iteration scale refers to the first method, GRASP-PR.

This allows observing an increase in time of a few percent for the execution of the path relinking method and a decrease for the fast ant system, because the solutions generated with the pheromone trails are closer to local optima, which speed up the descent method. This diagram presents a significantly different insight into the behavior of these methods. Indeed, a misinterpretation of Fig. 11.3 would suggest that up to 1000 iterations, the FANT method is better than GRASP-PR and, beyond that, the latter is the best. Repeating the improvement methods from random solutions is much worse.

Figure 11.5 shows that up to about 50 iterations, the 3 methods do not produce solutions of statistically different value. Only from 300 iterations onward can we clearly state that multiple descents are less efficient than GRASP-PR. The curves of the latter two methods cross at several points, but it is not possible to state that one produces solutions with a significantly lower average value than the other for a number of iterations less than 300. For more details on the generation of this type of diagrams, the reader can refer to [3] and to [1] for bootstrapping techniques.

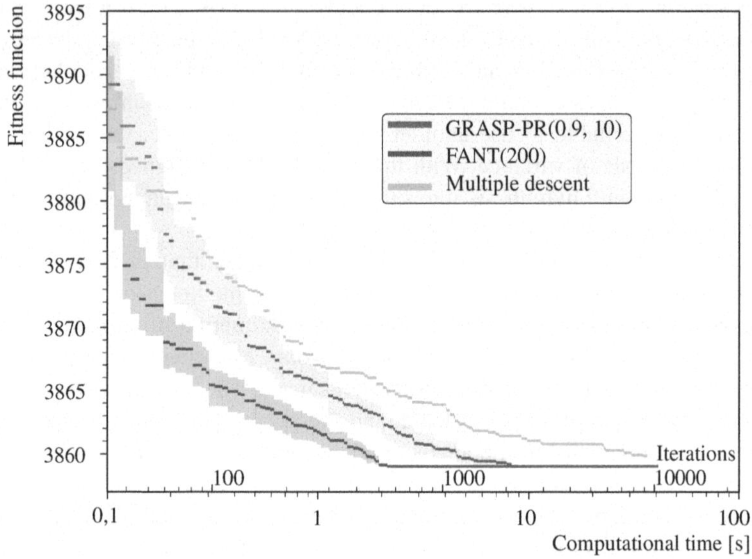

Fig. 11.5 Evolution of the average tour length as a function of the number of iterations for some codes presented in this book. Each method was run 20 times independently on the tsp225 instance. The shaded areas give the 95% confidence interval of the mean, obtained by exploiting a resampling technique

Problems

11.1 Data Structure Options for Multi-objective Optimization

It is argued in Sect. 5.5.3 that using a simple linked list to store the Pareto set may be inefficient. Is the more complicated KD-tree implementation really justified? To answer this question, evaluate the number of solutions produced by the Pareto local search Code 12.8, as well as the number of times one has to compare a neighbor solution to one of the solutions stored in this set. Deleting an element from a KD-tree can also be costly as a whole subtree has to be examined, and this can lead to cascading deletions. With a linked list, deleting a given element is done in constant time. Also assess the extra work involved.

11.2 Comparison of a True Simulated Annealing and a kind of SA with Systematic Neighborhood Evaluation

Compare the simulated annealing Code 7.1 and the noising method Code 7.2 when executed under the following conditions: instance with 50 cities and random distances generated uniformly between 1 and 99 (call to `rand_sym_matrix function`); start with a random solution (`rand_permutation` function); initial temperature: tour length/50; and final temperature: tour length/2500; $\alpha = 0.999$.

References

1. Bradley, E., Tibshirani, R.J.: An Introduction to the Bootstrap. Chapman and Hall (1994)
2. López-Ibáñez, M., Dubois-Lacoste, J., Pérez Cáceres, L., Birattari, M., Stützle, T.: The Irace package: iterated racing for Automatic Algorithm configuration. Oper. Res. Perspect. **3**, 43–58 (2016). https://doi.org/10.1016/j.orp.2016.09.002
3. Taillard, É.D.: Tutorial: Few guidelines for analyzing methods. In: Metaheuristic International Conference (MIC'05) Proceedings, Wien, Austria (2005)
4. Taillard, É.D., Waelti, P., Zuber, J.: Few statistical tests for proportions comparison. Eur. J. Oper. Res. **185**(3), 1336–1350 (2008). https://doi.org/10.1016/j.ejor.2006.03.070
5. Wolpert, D.H., Macready, W.G.: No free lunch theorems for optimization. IEEE Trans. Evol. Comput. **1**(1), 67–82 (1997). https://doi.org/10.1109/4235.585893

Chapter 12
Codes

This appendix provides the codes of utility procedures (random number generation, TSP data structures, KD-tree) appearing in various algorithms discussed in this book. Then, several codes are provided for testing complete metaheuristics.

These codes have been simplified in such a way that a user who is not familiar with coding or the use of program libraries can quickly execute them. Their programming style is therefore not so exemplary!

12.1 Random Numbers

Code 12.1 random_generators.py Implementation of a pseudo-random generator due to l'Ecuyer [1] as well as utility functions to generate integers between two bounds, random permutations and random symmetric matrices

```
######### L'Ecuyer random generator; retuns a value in ]0; 1[
def rando():
    m, m2 = 2147483647, 2145483479
    a12, a13, a21, a23 = 63308, -183326, 86098, -539608
    q12, q13, q21, q23 = 33921, 11714, 24919, 3976
    r12, r13, r21, r23 = 12979, 2883, 7417, 2071
    invm = 4.656612873077393e-10
    h = rando.x10 // q13
    p13 = -a13 * (rando.x10 - h * q13) - h * r13
    h = rando.x11 // q12
    p12 = a12 * (rando.x11 - h * q12) - h * r12
    if p13 < 0: p13 = p13 + m
    if p12 < 0: p12 = p12 + m
    rando.x10, rando.x11, rando.x12 = rando.x11, rando.x12, p12 - p13
    if rando.x12 < 0: rando.x12 = rando.x12 + m
    h = rando.x20 // q23
    p23 = -a23 * (rando.x20 - h * q23) - h * r23
```

Electronic Supplementary Material The online version of this article (https://doi.org/10.1007/978-3-031-13714-3_12) contains supplementary material, which is available to authorized users.

É. D. Taillard, *Design of Heuristic Algorithms for Hard Optimization*, Graduate Texts in Operations Research, https://doi.org/10.1007/978-3-031-13714-3_12

```
    h = rando.x22 // q21
    p21 = a21 * (rando.x22 - h * q21) - h * r21
    if p23 < 0: p23 = p23 + m2
    if p21 < 0: p21 = p21 + m2
    rando.x20, rando.x21, rando.x22 = rando.x21, rando.x22, p21 - p23
    if rando.x22 < 0: rando.x22 = rando.x22 + m2
    if rando.x12 < rando.x22: h = rando.x12 - rando.x22 + m
    else: h = rando.x12 - rando.x22
    if h == 0: return 0.5
    else: return h * invm

rando.x10, rando.x11, rando.x12 = 12345, 67890, 13579
rando.x20, rando.x21, rando.x22 = 24680, 98765, 43210

######## Returns a random integer in [low; high]
def unif(low, high):
    return low + int((high - low + 1) * rando())

######## Returns a random permutation of the elements 0...n-1
def rand_permutation(n):
    p = [i for i in range(n)]
    for i in range(n - 1):
        random_index = unif(i, n - 1)
        p[i], p[random_index] = p[random_index], p[i]
    return p

######### Returns a symmetric n x n matrix of random numbers with 0 diagonal
def rand_sym_matrix(n, low, high):
    matrix = [[0 for _ in range(n)] for _ in range(n)]
    for i in range(n-1):
        for j in range(i+1, n):
            matrix[i][j] = matrix[j][i] = unif(low, high)
    return matrix
```

12.2 TSP Utilities

Code 12.2 tsp_utilities.py Utility functions for the traveling salesman: computation of the length of a tour when a solution is provided in the form of an array giving the order in which the cities are visited; transformation of a solution from one form to another (order, successors, predecessors); comparison of tours

```
######### Compute the length of a TSP tour
def tsp_length(d,                                          # Distance matrix
               tour):                                      # Order of the cities
    n = len(tour)
    length = d[tour[n - 1]][tour[0]]
    for i in range(n - 1):
        length += d[tour[i]][tour[i + 1]]
    return length

######### Build solution representation by successors
def tsp_tour_to_succ(tour):
    n = len(tour)
    succ = [-1] * n
    for i in range(n):
        succ[tour[i]] = tour[(i+1)%n]
    return succ

######### Build solution representation by predeccessors
def tsp_succ_to_pred(succ):
    n = len(succ)
```

```
    pred = [-1] * n
    for i in range(n):
        pred[succ[i]] = i
    return pred

######### Convert solution from successor of each city to city order
def tsp_succ_to_tour(succ):
    n = len(succ)
    tour = [-1] * n
    j = 0
    for i in range(n):
        tour[i] = j
        j = succ[j]
    return tour

######### Convert a solution given by 2-opt data structure to a standard tour
def tsp_2opt_data_structure_to_tour(t):
    n = int(len(t)/2 + 0.5)
    tour = [-1] * n
    j = 0
    for i in range(n):
        tour[i] = j>>1
        j = t[j]
    return tour

######### Compare 2 directed tours; returns the number of different arcs
def tsp_compare(succ_a, succ_b):
    n = len(succ_a)
    count = 0
    for i in range(n):
        if succ_a[i] != succ_b[i]:
            count += 1
    return count
```

12.3 TSP Lin and Kernighan Improvement Procedure

Code 12.3 tsp_ LK.py Ejection chain for the TSP

```
from tsp_utilities import *                                        # Listing 12.2

######### Basic Lin & Kernighan improvement procedure for the TSP
def tsp_LK(D,                                                    # Distance matrix
           tour,                                                         # Solution
           length):                                                   # Tour length
    n = len(tour)
    succ = tsp_tour_to_succ(tour)
    for i in range(n): succ[tour[i]] = tour[(i + 1) % n]
    tabu = [[0 for _ in range(n)] for _ in range(n)] #Can edge i-j be removed ?
    iteration = 0              # Outermost loop counter to identify tabu condition
    last_a, a = 0, 0                     # Initiate ejection chain from city a = 0
    improved = True
    while a != last_a or improved:
        improved = False
        iteration += 1
        b = succ[a]
        path_length = length - D[a][b]
        path_modified = True
```

```
        while path_modified: # Identify best ref. struct. with edge a-b removed
            path_modified = False
            ref_struct_cost = length     # Cost of reference structure retained
            c = best_c = succ[b]
            while succ[c] != a:                    # Ejection can be propagated
                d = succ[c]
                if path_length - D[c][d] + D[c][a] + D[b][d] < length:
                    best_c = c              # An improving solution is identified
                    ref_struct_cost = path_length - D[c][d] + D[c][a] + D[b][d]
                    break                 # Change improving solution immediately
                if tabu[c][d] != iteration and \
                                path_length + D[b][d] < ref_struct_cost:
                    ref_struct_cost = path_length + D[b][d]
                    best_c = c
                c = d                                    # Next value for c and d
            if ref_struct_cost < length: # Admissible reference structure found
                path_modified = True
                c, d = best_c, succ[best_c]         # Update reference structure
                tabu[c][d] = tabu[d][c] = iteration#Don't remove again edge c-d
                path_length += (D[b][d] - D[c][d])
                i, si, succ[b] = b, succ[b], d              # Reverse path b -> c
                while i != c:
                    succ[si], i, si = i, si, succ[si]
                b = c

                if path_length + D[a][b] < length: # A better solution is found
                    length = path_length + D[a][b]
                    succ[a], last_a = b, b
                    improved = True
                    tour = tsp_succ_to_tour(succ)
        succ = tsp_tour_to_succ(tour)
        a = succ[a]
    return tour, length
```

12.4 KD-Tree Insertion and Inspection

Code 12.4 kd_tree_add_scan.py Codes to define the general structure of the nodes of a KD-tree, to add an element to a KD-tree, and to inspect the whole tree. The inspection procedure just prints out the elements

```
######### KD tree node data structure
class Node:

    def __init__(self, key, father, info):                # Create a tree node
        self.key = key[:]                          # Key used to separate nodes
        self.father = father    # Father of the node, (None if node is the root)
        self.info = info[:]             # Information to store in the node (list)
        self.left = None                                        # Left son of node
        self.right = None                                              # Right son

K = 3                                     # Define the dimension of the KD tree

####################### Add a new node in a KD tree ##########################
def kd_tree_add(root,                                       # Root of a (sub-) tree
                key,                                      # Key for splitting nodes
                info,                               # Information to store in the node
                depth):                                          # Depth of the root
```

```

    if root is None:
        root = Node(key, None, info)
    elif root.key[depth % K] < key[depth % K]:
        if root.right is None:
            root.right = Node(key, root, info)
        else:
            root.right = kd_tree_add(root.right, key, info, depth + 1)
    else:
        if root.left is None:
            root.left = Node(key, root, info)
        else:
            root.left = kd_tree_add(root.left, key, info, depth + 1)

    return root

############# Scan a KD tree and print key and info for each node ############
def kd_tree_scan(root):

    if root:
        if root.left:
            kd_tree_scan(root.left)
        print('Key: ', root.key, ' Info: ', root.info)
        if root.right:
            kd_tree_scan(root.right)
```

12.5 KD-Tree Delete

Code 12.5 kd_tree_delete.py Code for removing a node in a KD-tree

```
########## Find the node with min or max value in a given dimension ##########
def kd_tree_find_opt(root,                            # Root of a KD (sub-)tree
                     dim,              # Dimension in which optimum is looked for
                     depth,                                    # Depth of the root
                     minimum,         # Look for minimum (True) or maximum (False)
                     value,                             # Best value already known
                     opt):                              # Node with optimum value
    depth_opt = -1
    if (minimum and (value > root.key[dim])) \
        or ((not minimum) and (value < root.key[dim])):
        opt = root
        value = root.key[dim]
        depth_opt = depth
    if root.left:
        opt, value, depth_opt = kd_tree_find_opt(root.left, dim, depth + 1,
                                                 minimum, value, opt)
    if root.right:
        opt, value, depth_opt = kd_tree_find_opt(root.right, dim, depth + 1,
                                                 minimum, value, opt)
    return opt, value, depth_opt

################### Delete the root of a KD (sub-) tree ######################
def kd_tree_delete(root,                                         # Node to delete
                   depth):                                     # Depth of the node
    from kd_tree_add_scan import K                                   #Listing 12.4
    if root.left:       # Root is an internal node, must be replaced
        replacing, val_repl, depth_repl = kd_tree_find_opt(root.left,
```

```
                                    depth % K, depth + 1, False, float('-inf'), None)
    elif root.right:
        replacing, val_repl, depth_repl  = kd_tree_find_opt(root.right,
                                      depth % K, depth + 1, True, float('inf'), None)
    else:                                                          # The node is a leaf
        if root.father:
            if root.father.left == root:
                root.father.left = None
            else:
                root.father.right = None
        return None                                        # A leaf is directly deleted

    root.key = replacing.key[:]
    root.info = replacing.info
    kd_tree_delete(replacing, depth_repl)
    return root
```

12.6 KD-Tree Update Pareto Set

Code 12.6 kd_tree_update_pareto.py Code for updating a Pareto set represented by a KD-tree

```
from kd_tree_add_scan import K, kd_tree_add                            #Listing 12.4

########## Tell if node is in the box bounded by minimum and maximum ##########
def kd_tree_in(node,
               minimum, maximum):            # Lower and upper corner of the box
    i, result = 0, True
    while i < K and result:
        result = minimum[i] <= node.key[i] <= maximum[i]
        i = i + 1
    return result

### Find a node (if any) with its depth in the box bounded by mini and maxi ###
def kd_tree_find(root,         # Root of the tree in which the node is looked for
                 mini, maxi,                 # Lower and upper corner of the box
                 depth):                                     # Depth of the root
    if root is None:
        return None, -1
    if kd_tree_in(root, mini, maxi):
        return root, depth

    if maxi[depth%K] >= root.key[depth%K]:
        result, depth_found = kd_tree_find(root.right, mini, maxi, depth + 1)
        if result:
            return result, depth_found
    if mini[depth%K] <= root.key[depth%K]:
        result, depth_found = kd_tree_find(root.left, mini, maxi, depth + 1)
        if result:
            return result, depth_found
    return None, -1

########## Remove points of Pareto front dominated by costs (if any) ##########
def update_3opt_pareto(pareto,                     # Current Pareto front to update
                       costs,                     # New point to be eventually added
                       successors, distances):       # Problem solution and data
    from kd_tree_delete import kd_tree_delete                          #Listing 12.5
    from tsp_3opt_pareto import tsp_3opt_pareto                        #Listing  5.5
```

```
    minimum = [0 for _ in range(K)]
    maximum = [float('inf') for _ in range(K)]
    dominant, depth = kd_tree_find(pareto, minimum, costs, 0)
    if dominant is None:                    # No point of pareto dominates costs
        while True:           # There are dominated points, costs improves pareto
            dominated, depth = kd_tree_find(pareto, costs, maximum, 0)
            if dominated is None:                    # All dominated points removed
                break
            if dominated == pareto:
                pareto = kd_tree_delete(dominated, depth)
            else:
                dominated = kd_tree_delete(dominated, depth)

        pareto = kd_tree_add(pareto, costs, successors, 0)
        pareto = tsp_3opt_pareto(pareto, costs, successors, distances)
    return pareto
```

12.7 TSP 2-Opt and 3-Opt Test Program

Code 12.7 test_tsp_2_and_3opt.py This code first generates a symmetric matrix with random distances and starts with a random solution. The latter is improved with a local search applying the first-move improving policy with the 2-opt neighborhood. This method is relatively rapid for instances with up to a few thousand cities. Then, all sub-paths of 100 successive cities in this solution are improved with a 3-opt neighborhood. This method runs in almost linear time, but only produces good solutions if the starting solution is adequate. The solution is then improved with a full 3-opt neighborhood. Its complexity is considerably higher; the computational time becomes significant beyond a few hundred cities. Ultimately, the solution is improved with the 2-opt neighborhood, but applying the best move policy at each iteration

```
'''
Programme to test various local improvement methods
Example of execution:
Number of cities:
500
Random solution: 24565
Cost of solution found with 2-opt first 953
Solution improved with 3-opt limited (100 cities) 738
Solution improved with complete 3-opt 679
Solution improved with 2-opt best 673
'''
import math

from random_generators import *                              # Listing 12.1
from tsp_utilities import *                                  # Listing 12.2
from tsp_2opt_first import tsp_2opt_first                    # Listing  5.4
from tsp_2opt_best import tsp_2opt_best                      # Listing  5.1
from tsp_3opt_limited import tsp_3opt_limited                # Listing  6.1
from tsp_3opt_first import tsp_3opt_first                    # Listing  5.2

print('Number of cities: ')
n = int(input())

distances = rand_sym_matrix(n, 1, 99)
solution = rand_permutation(n)

length = tsp_length(distances, solution)
```

```
print('Random solution: {:d}'.format(length))

solution, length = tsp_2opt_first(distances, solution, length)
print('Cost of solution found with 2-opt first: {:d}'.format(length))

successors = tsp_tour_to_succ(solution)
successors, length = tsp_3opt_limited(distances, 100, successors, length)
print('Solution improved with 3-opt limited (100 cities): {:d}'
      .format(length))

successors, length = tsp_3opt_first(distances, successors, length)
print('Solution improved with complete 3-opt: {:d}'.format(length))

solution = tsp_succ_to_tour(successors)
solution, length = tsp_2opt_best(distances, solution, length)
print('Solution improved with 2-opt best: {:d}'.format(length))
```

12.8 Multi-objective TSP Test Program

Code 12.8 test_tsp_3opt_pareto.py A program for testing a local Pareto search for a TSP with a randomly generated distance matrix. For a 20-city and 3-objective instance, this program generates an approximation to the Pareto set with more than 3000 solutions. Since the implementation is highly recursive, the recursion stack and console user limits must be appropriately resized

```
'''
Programme to test pareto local seach for the TSP with 3-opt moves
Example of run with K = 3 (KD-tree Key = costs; Info = tour)

Number of cities:
6
Key:  [137, 253, 273]  Info:  [5, 3, 0, 4, 2, 1]
Key:  [173, 287, 236]  Info:  [2, 4, 5, 0, 3, 1]
Key:  [222, 288, 235]  Info:  [2, 3, 4, 0, 5, 1]
Key:  [263, 249, 242]  Info:  [3, 5, 4, 1, 0, 2]
Key:  [172, 265, 244]  Info:  [4, 5, 0, 1, 3, 2]
Key:  [182, 224, 320]  Info:  [5, 2, 4, 0, 3, 1]
Key:  [184, 244, 297]  Info:  [1, 5, 4, 0, 3, 2]
Key:  [166, 340, 264]  Info:  [1, 3, 0, 4, 5, 2]
Key:  [246, 367, 201]  Info:  [3, 4, 0, 1, 5, 2]
'''
import sys
from random_generators import rand_sym_matrix                    # Listing 12.1
from kd_tree_add_scan import K, kd_tree_scan                     # Listing 12.4
from tsp_3opt_pareto import tsp_3opt_pareto                      # Listing  5.5

sys.setrecursionlimit(50000)   # Higly recursive implementation; enlarge stack!
print('Number of cities: ')
n = int(input())

distance = [rand_sym_matrix(n,1,99) for _ in range(K)]
successors = [(i + 1) % n for i in range(n)]                 # Initial solution
costs = [0 for _ in range(K)]
for dim in range(K):
    for i in range(n):
        costs[dim] += distance[dim][i][successors[i]]

```

```
pareto = tsp_3opt_pareto(None, costs, successors, distance)

kd_tree_scan(pareto) #Print pareto front with tours (successors representation)
```

12.9 Fast Ant TSP Test Program

Code 12.9 test_tsp_FANT.py Program to test a method inspired by artificial ant colonies

```
'''
Programme to test the Fast Ant procedure
Example of execution

Number of cities:
200
Number of FANT iterations:
200
FANT parameter:
30
FANT 1  314
FANT 2  310
FANT 75  308
FANT 175  306
Cost of solution found with FANT 306
'''

from random_generators import rand_sym_matrix                          # Listing 12.1
from tsp_FANT import tsp_FANT                                          # Listing  8.3

print('Number of cities: ')
n = int(input())
print('Number of FANT iterations: ')
fant_iterations = int(input())
print('FANT parameter (best solution reinforcement): ')
fant_parameter = int(input())


distances = rand_sym_matrix(n, 1, 99)
tour, cost = tsp_FANT(distances, fant_parameter, fant_iterations)
print('Cost of solution found with FANT: {:d}'.format(cost))
```

12.10 Taboo Search TSP Test Program

Code 12.10 test_tsp_TS.py A taboo search test program for a TSP with a randomly generated symmetric distance matrix

```
'''
Programme to test a Taboo Search for the TSP
Example of run:
Number of cities:
30
Number of tabu iterations:
200
Minimum tabu_duration:
4
```

```
Maximum tabu_duration:
20
Penalty:
0.005
TS 1 1190
TS 2 1053
TS 3 906
TS 4 796
...
TS 29 177
TS 46 174
TS 119 173
Cost of solution found : 173
[13, 0, 6, 5, 27, 28, 1, 4, 29, 20, 2, 7, 16, ...24, 9, 21, 25, 10, 22, 19, 14]
'''
import math

from random_generators import *                                    # Listing 12.1
from tsp_utilities import tsp_length                               # Listing 12.2
from tsp_TS import tsp_TS                                          # Listing  9.1

print('Number of cities: ')
n = int(input())

distances = rand_sym_matrix(n, 1, 99)
solution = rand_permutation(n)
length = tsp_length(distances, solution)


print('Number of tabu iterations: ')
iterations = int(input())
print('Minimum tabu_duration: ')
min_tabu = int(input())
print('Maximum tabu_duration: ')
max_tabu = int(input())
print('Penalty: ')
freq_penalty = float(input())
solution, length = tsp_TS(distances, solution, length,
                          iterations, min_tabu, max_tabu, freq_penalty)
print('Cost of solution found : {:d}'.format(length))
print(solution)
```

12.11 Memetic TSP Test Program

Code 12.11 test_tsp_GA.py A memetic algorithm test program for a TSP with a randomly generated symmetric distance matrix

```
'''
Programme to test a basic memetic algorithm
Example of execution:
Number of cities:
200
Size of the population:
10
Mutation rate:
0.02
Number of generations:
```

```
30
GA initial best individual 9424
GA improved tour 0 313
GA improved tour 7 312
GA improved tour 15 307
Cost of solution found with GA: 307

'''


from random_generators import rand_sym_matrix                    # Listing 12.1
from tsp_GA import tsp_GA                                         # Listing 10.5

print('Number of cities: ')
n = int(input())
print('Size of the population: ')
population_size = int(input())
print('Mutation rate: ')
mutation_rate = float(input())
print('Number of generations: ')
nr_generations = int(input())

distances = rand_sym_matrix(n, 1, 99)
_, cost = tsp_GA(distances, population_size, nr_generations, mutation_rate)
print('Cost of solution found with GA: {:d}'.format(cost))
```

12.12 GRASP with Path Relinking TSP Test Program

Code 12.12 test_tsp_GRASP_PR.py A GRASP with path relinking test program. This method uses GRASP which calls for a local search based on ejection chains, as well as other utility functions

```
'''
Programme to test a GRASP with Path Relinking
Example of execution:
Number of cities:
200
Iterations:
50
Population size:
10
Alpha:
0.7
GRASP_PR population updated: 11 319
GRASP_PR population updated: 12 317
GRASP_PR population updated: 13 318
GRASP_PR population updated: 15 318
GRASP_PR population updated: 16 320
GRASP_PR population updated: 18 317
GRASP_PR population updated: 19 315
GRASP_PR population updated: 20 316
GRASP_PR population updated: 21 309
GRASP_PR population updated: 25 316
GRASP_PR population updated: 26 313
GRASP_PR population updated: 35 316
GRASP_PR population updated: 36 313
GRASP_PR population updated: 40 313
```

```
GRASP_PR population updated: 42 311
GRASP_PR population updated: 49 313
Cost of solution found with GRASP_PR: 309
'''
from random_generators import rand_sym_matrix                    # Listing 12.1
from tsp_GRASP_PR import tsp_GRASP_PR                            # Listing 10.7

print('Number of cities: ')
n = int(input())
print('Iterations: ')
iterations = int(input())
print('Population size: ')
population_size = int(input())
print('Alpha: ')
alpha = float(input())

distances = rand_sym_matrix(n, 1, 99)

tour, length = tsp_GRASP_PR(distances, iterations, population_size, alpha)
print('Cost of solution found with GRASP_PR: {:d}'.format(length))
```

References

1. L'Ecuyer P.: Combined multiple recursive random number generators. Operations Research **44**(5), 816–822 (1996). http://www.jstor.org/stable/171570

Solutions to the Exercises

Problems of Chap. 1

1.1 Draw 5 Segments

A common mistake, when trying to solve this type of problem, is to take a sheet of paper and start drawing segments more or less at random. After unsuccessfully scribbling for a few minutes, we look for a more systematic proof.

For instance, without loss of generality, we can assume that segment 1 cuts segments 2, 3, and 4. So, segment 1 does not cut segment 5. This implies that segment 5 necessarily cuts segments 2, 3, and 4. To complete the crossings, if we assume that segment 2 intersects segment 3, there is no longer any possibility for segment 4 to cut other segments. We can deduce, by enumeration of all the other possible hypotheses, that the problem has no solution. This way of proceeding is not reasonable if we want to show it is impossible to have 301 segments which each intersects 3 others: the combinatorics is such that we will never arrive at the end of the demonstration.

A natural graph model—a crossing $\equiv$ a vertex; a segment $\equiv$ an edge—does not lead to something productive. In contrast, if a vertex corresponds to a segment and an edge represents the relationship that two segments intersect, then the solution of the problem becomes obvious. We are looking for a graph with five vertices of degree 3. So, we are looking for a graph whose sum of degrees is equal to $5 \cdot 3 = 15$. Since the sum of the degrees is even in any graph, we deduce the impossibility of drawing 5 (or 301) segments such that each one crosses three others.

É. D. Taillard, *Design of Heuristic Algorithms for Hard Optimization*, Graduate Texts in Operations Research, https://doi.org/10.1007/978-3-031-13714-3

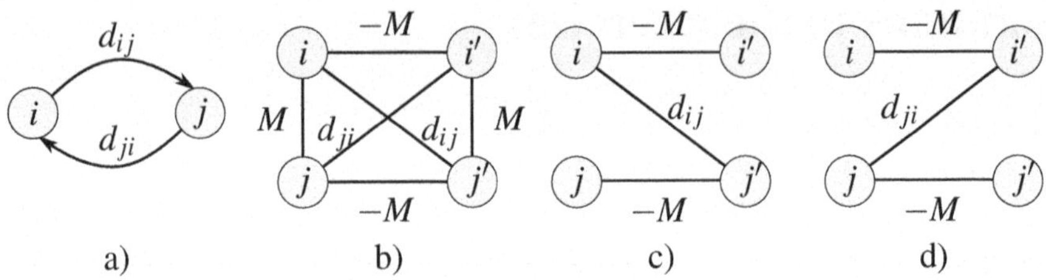

Fig. 1 Principle of a polynomial transformation of an asymmetric traveling salesman problem into a symmetric one. (**a**) Two nodes i and j of the original directed graph. (**b**) Doubling of nodes and weights in an undirected graph; M is a sufficiently large constant value. (**c**) A possibility of visiting vertices i and j in the undirected graph, without having to pay the M penalty and by collecting a bonus of $2M$, corresponding to a visit in the order $i \rightarrow j$ in the directed graph. (**d**) The only other reasonable possibility, corresponding to a visit in the order $j \rightarrow i$

1.2 O Simplification

- $O(2^n)$
- $O(2^{2^n})$
- $\Omega((n+2)!)$
- $\Omega(n \log(\log(n)))$
- $O(n^{\log(n)})$
- $O(n^5)$

1.3 Turing Machine Program

The transition function δ is given by the following table:

	Symbol on the tape (Γ)				
State (Q)	b	a	c	e	n
q_0	q_N	$(q_a, a, 1)$	$(q_0, c, 1)$	$(q_0, e, 1)$	$(q_0, n, 1)$
q_a	q_N	$(q_a, a, 1)$	$(q_0, c, 1)$	$(q_0, e, 1)$	$(q_{an}, n, 1)$
q_{an}	q_N	$(q_a, a, 1)$	$(q_0, c, 1)$	q_Y	$(q_0, n, 1)$

1.4 Clique is NP-Complete

Section 1.2.3.4 proofs that finding a stable set of a given size is NP-Complete. In the complementary graph, this problem is equivalent to looking for a clique of this size. Therefore, any stable set instance transforms polynomially into a clique instance.

1.5 Asymmetric TSP to Symmetric TSP

See Fig. 1.

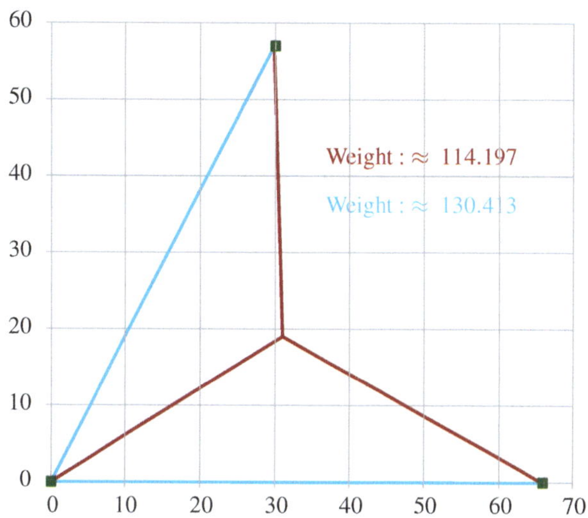

Fig. 2 For the numerical application, a Steiner point at approximate coordinates (31, 19) is necessary (so that the three segments issuing from this point toward the vertices of the triangle make angles of 120°). The Steiner tree thus constructed has a weight of about 114.197, while that of the minimum spanning tree is about 130.413

Problems of Chap. 2

2.1 Connecting Points

The problem here has not been formulated very precisely. This is a recurrent issue between two interlocutors who do not have the same background. For example, the manager of a company trying to explain the functionality of an application to a programmer. One forgets to mention certain constraints that seem obvious, and the other tries to transcribe the problem into a known algorithm but which is not satisfactorily modeling the reality. In this exercise, it has not been specified if the connections between the vertices had to be straight lines. If so, the solution consists in drawing a minimum spanning tree and the problem is simple. If the connection between two vertices is not necessarily a unique straight line segment, then the problem is to seek a Steiner tree and the problem is intractable. See Fig. 2.

2.2 Accessibility by Lorries

This is to maximize the minimum cut from A to B. The problem can be solved by calculating a maximum weight spanning tree in the network. The unique chain connecting A and B in this tree has the highest possible cut. For the numerical application, the chain is $A - 5 - 1 - 7 - 4 - 3 - 8 - B$ and the cut has a value of 48.

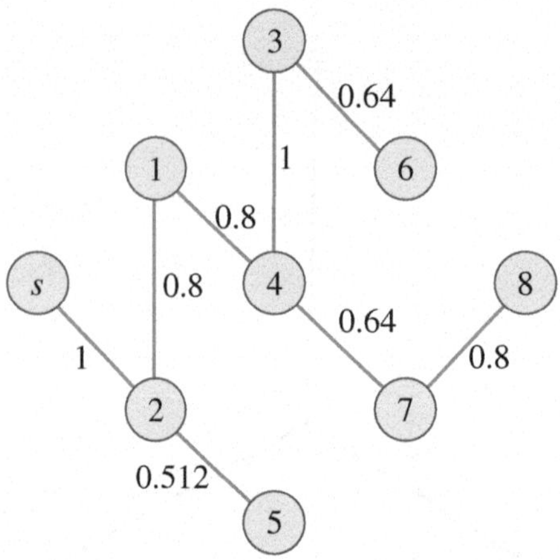

Fig. 3 Connections to keep

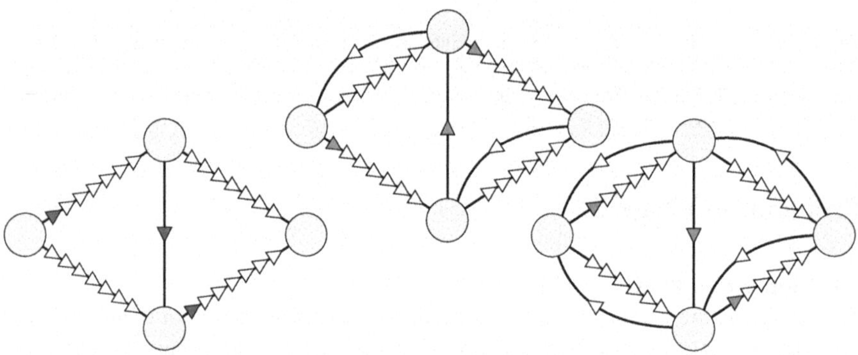

Fig. 4 Successive residue networks when applying the Ford and Fulkerson algorithm. An arc of very low capacity can be alternately used in the normal direction and then its flow canceled at the next iteration

2.3 Network Reliability

Since the weights correspond to probabilities that must be multiplied and not added, a standard algorithm of path length minimization can be used by taking the opposite of the logarithm of the probabilities. See Fig. 3.

2.4 Ford & Fulkerson Algorithm Degeneracy

The number of iterations can grow up to the ratio that exists between the arc with the largest capacity and the one that has the lowest. See Fig. 4.

2.5 TSP Permutation Model
Problem instance data: $n \times n$ distance matrix $D = (d_{ij})$. Objective: find a permutation p minimizing:

$$\sum_{i=1}^{n-1} d_{p_i p_{i+1}} + d_{p_n p_1}$$

2.6 PAM and k-Means Implementation
The algorithmic complexity of these two procedures can hardly be expressed theoretically, because the loops in lines 9 and 6 are repeated an indeterminate number of times. In practice, we observe a number of repetitions more or less proportional to k for Algorithm 2.7. This number is much lower for Algorithm 2.8, typically less than 20. More precisely, the number of repetitions for Algorithm 2.7 increases very weakly with n when k is constant (we observe an increase from $O(n^{0.2})$ to $O(n^{0.3})$, approximately), while this increase is approximately linear for $k = n/20$.

Although very often used in practice, the k-means algorithm does not produce good solutions if given random initial center positions. Starting from the solution provided by Algorithm 2.7 gives much better results. However, the computational time to obtain this solution can be prohibitive.

2.7 Optimality Criterion
The scheduling is optimal because the last machine has to wait the shortest possible time before starting to work, and then it works continuously until the end.

2.8 Flowshop Makespan Evaluation
The earliest ending time for processing the ith object (p_i) on machine j is given with the recurrence relation:

$$f_{ij} = \begin{cases} 0 & \text{If } i = 0 \text{ or if } j = 0 \\ \max(f_{i-1j}, f_{ij-1}) + t_{p_i j} & \text{Otherwise} \end{cases}$$

The latest starting time of this operation is given with the recurrence relation:

$$d_{ij} = \begin{cases} f_{mn} & \text{If } i = n+1 \text{ or if } j = m+1 \\ \min(d_{i+1j}, d_{ij+1}) - t_{p_i j} & \text{Otherwise} \end{cases}$$

2.9 Transforming the Knapsack Problem into the Generalized Assignment
Let I be the set of n objects of revenue c_i and volume v_i and V the volume of the knapsack. We can create a generalized assignment problem with a set U consisting of $m = 2$ elements. The element $u = 1$ corresponds to the objects that must be put in the knapsack and the element $u = 2$ to those that remain outside. By solving a generalized assignment problem with $c_{i1} = 0$, $c_{i2} = c_i$, $w_{i1} = v_i$, $w_{i2} = 0$, $t_1 = V$ and $t_2 = 0$, one minimizes the value of the objects remaining outside the knapsack while satisfying the volume constraint for those put in the knapsack. Therefore, knowing that the knapsack is NP-hard, this proves that the generalized assignment problem is also NP-hard.

Problems of Chap. 3

3.1 Assigning Projects to Students

The assignment problem can be solved by finding a maximum flow with minimum cost in a bipartite graph. See Fig. 5.

- First step: *a* takes 1.
- Second step: *c* takes 2.
- Third step: *a* changes and takes 3; *b* takes 1.
- Last step: *a* changes again and takes 4; *d* takes 3.

3.2 Placing Production Units

Consider the distance matrix D, flow matrix F, and assignment costs $D \times F$ with:

$D =$

	1	2	3
a	4	5	5
b	7	2	2
c	5	4	6

$F =$

	4	5	6
1	7	3	1
2	3	8	6
3	2	1	9

$D \times F =$

	4	5	6
a	**53**	57	79
b	59	39	**37**
c	59	**53**	83

Numerical application: a—4, b—6, c—5.

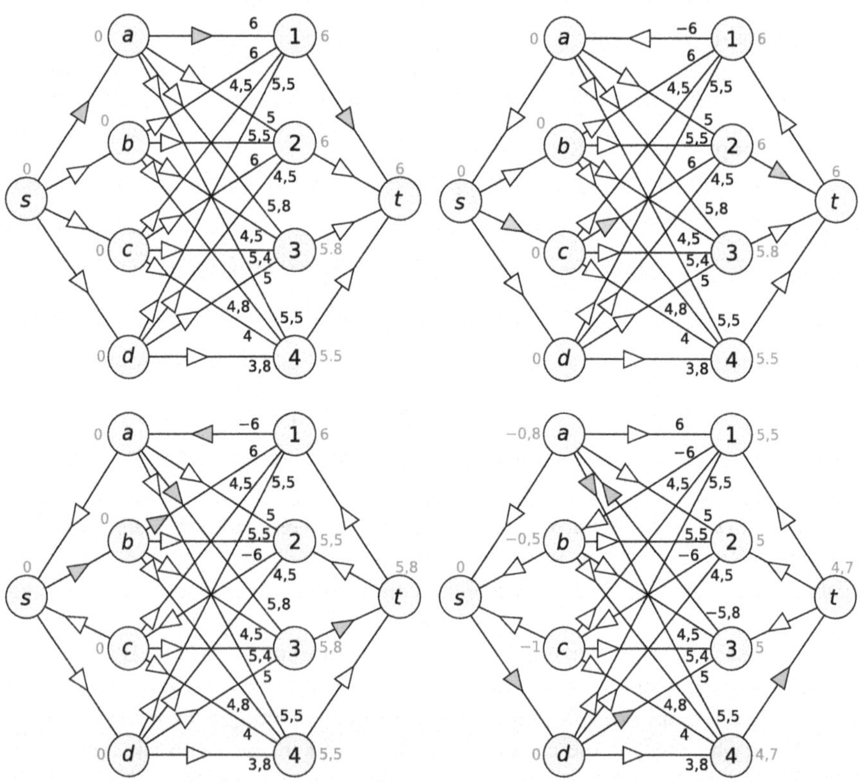

Fig. 5 The successive flow increases for the optimal assignment of the four projects

If there are flows between the new units, the problem transforms into a quadratic assignment which is NP-hard.

3.3 Oral Examination

The problem can be modeled by coloring the edges of a bipartite graph. To ensure we can decompose the graph into a minimum number of perfect matchings, we must complete it so that there is the same number of modules than students. If there are fewer modules than students, create dummy modules. Next, we add dummy edges so that each vertex has the same degree. Thus, after finding a first matching, the corresponding edges can be removed. We can start again with a graph possessing the same properties.

3.4 Written Examination

The problem can be modeled by coloring the vertices of a graph. The vertices correspond to the modules to be examined. The edges correspond to incompatibilities between modules. All the vertices-modules a student must attend are completely connected by a clique. The problem is intractable. For the numerical example, 4-day timetables exist. For instance, $\{1, 7\}$, $\{2, 5\}$, $\{3, 6\}$, $\{4, 8\}$.

3.5 QAP with More Positions Than Items

If there are fewer elements (n) to place than positions (m), we can come back to the standard case with two $m \times m$ matrices by adding $m - n$ dummy elements with a zero flow between them.

If there is a fixed cost c_{ir} for assigning element i to position r, the objective must be changed:

$$\sum_{i=1}^{n} \sum_{j=1}^{n} f_{ij} d_{p_i p_j} + \sum_{i=1}^{n} c_{p_i}$$

3.6 Mobile Phone Keyboard Layout

The problem can be modeled by a QAP. As we only have 27 symbols to place for 36 positions, we extend the frequency matrix to have a 36×36 matrix. Let us consider the sub-matrices:

$$A = \begin{Bmatrix} 2\ 3\ 4\ 5 \\ 2\ 3\ 4\ 5 \\ 2\ 3\ 4\ 5 \\ 2\ 3\ 4\ 5 \end{Bmatrix} \text{ and } B = \begin{Bmatrix} 7\ 8\ 9\ 10 \\ 7\ 8\ 9\ 10 \\ 7\ 8\ 9\ 10 \\ 7\ 8\ 9\ 10 \end{Bmatrix}$$

The "distance" matrix (corresponding to times) is given by:

$$D = \begin{Bmatrix} B & A & \cdots & A \\ A & B & \cdots & A \\ \vdots & \vdots & \ddots & \vdots \\ A & A & \cdots & B \end{Bmatrix}$$

3.7 Graph Bipartition to QAP

Let A be the adjacency matrix of the graph (with $2n$ vertices). Let $\mathbf{0}$ and $\mathbf{1}$ two $n \times n$ matrices containing only 0s and 1s. The QAP instance with flow matrix given by A and distance matrix given by $\begin{Bmatrix} \mathbf{0} & \mathbf{1} \\ \mathbf{1} & \mathbf{0} \end{Bmatrix}$ is equivalent to the graph bipartition problem.

3.8 TSP to QAP

A TSP instance with a distance matrix D can be transformed into a QAP one with the same distance matrix and the flow matrix given by:

$$f_{ij} = \begin{cases} 1 \text{ If } j = i + 1 \text{ or if } i = n \text{ and } j = 1 \\ 0 \text{ Otherwise} \end{cases}$$

3.9 Special Bipartition

A binary vector can model a solution s and an example of a fitness function is:

$$\left(1170 - \sum_{i=1}^{50} i \cdot (1 - s_i)\right)^2 + \left|36{,}000 - \prod_{i=1}^{50} i \cdot s_i\right|$$

3.10 Magic Square

We could model a solution attempt for a magic square of order n by a permutation of the elements from 1 to n^2. A fitness function could be to sum the squares of the deviations from the target sum. But why on earth design a heuristic method for this simple problem? Indeed, polynomial algorithms exist for the construction of magic squares (except for $n = 2$).

3.11 Glass Plate Manufacturing

The no-wait flowshop sequencing problem can be modeled as an asymmetric TSP as follows. A fictitious plate is included with zero processing time on all the machines. The minimum difference between the starting time of the plate i and that of the plate k is given by:

$$d_{ik} = \max_r \left(\sum_{j=1}^{r} t_{ij} - \sum_{j=1}^{r-1} t_{kj} \right)$$

The (d_{ik}) matrix corresponds to the TSP distances. The optimum tour length corresponds to the minimum production time of the plates. The order of production is the same as that of the tour, taking care to start with the fictitious plate/city.

3.12 Optimal 1-tree

Choosing $\lambda_1 = 0, \lambda_2 = 0, \lambda_3 = 6, \lambda_4 = -4$, and $\lambda_5 = -2$ provides a 1-tree of weight 74. This 1-tree corresponds to the circuit $1 - 2 - 4 - 3 - 5 - 1$ which is therefore the optimal TSP tour for this instance.

Problems of Chap. 4

4.1 Random Permutation

The more the number of selected elements increases, the more unnecessary random draws must be made with Algorithm 4.5. At the last iteration, n trials are made on average while there is no alternative. Algorithm 4.6 can be implemented in $\Theta(n)$, but the permutations are not uniformly drawn.

An efficient algorithm for generating random permutation, evenly distributed, is given in Code 12.1. The operating principle is as follows. All the items are introduced in an array p. At the ith step, the array contains random items until the index i. Beyond this index are the items remaining to be chosen.

We can check that a given item has the same probability of being in any place in the array p. Trivially, it has a probability of $1/n$ to appear in the first place. The probability of appearing in second place is calculated by considering that it should not be chosen for the first place (probability $(n-1)/n$) and that it is chosen for the second (probability $1/(n-1)$), i.e. $(n-1)/n \cdot 1/(n-1) = 1/n$; etc.

4.2 Greedy Algorithms for the Knapsack

The incremental cost of adding an item to the knapsack can be defined by:

- The inverse of its revenue
- Its volume
- Its volume/revenue ratio

4.3 Greedy Algorithm for the TSP on the Delaunay

If the points are collinear, the Delaunay is a chain. In this case, we cannot build a tour. Figure 6 shows a nondegenerate Delaunay Triangulation with no Hamiltonian tour.

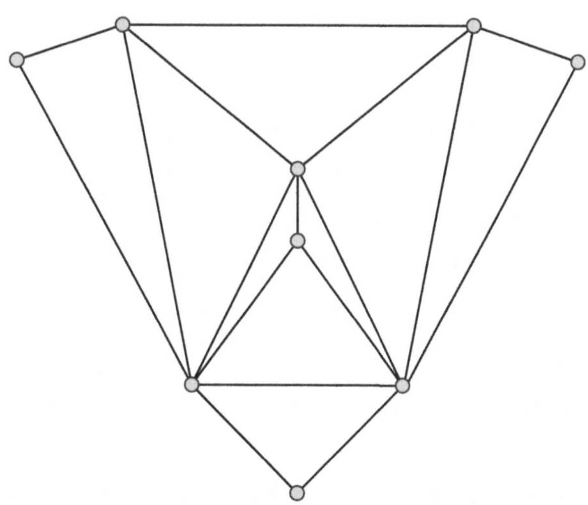

Fig. 6 A non-Hamiltonian Delaunay triangulation

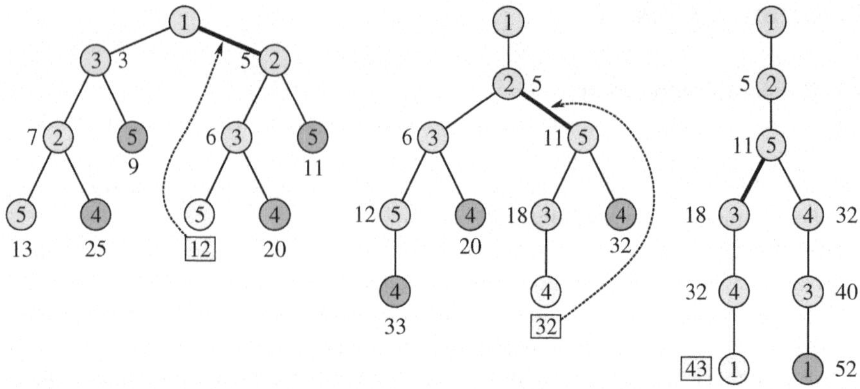

Fig. 7 Applying the beam search procedure to a TSP instance. At each level, both best partial solutions are retained and the tree is developed up to three levels

4.4 TSP with Edge Subset

The construction of a tour can effectively be completed in linear time if each city is only connected to its 40 nearest ones. However, these cities cannot be obtained in linear time. In addition, all the cities the closest to the last visited may already belong to the tour under construction.

4.5 Constructive Methods Complexity

The nearest neighbor heuristic for the TSP can be implemented in $\Theta(n^2)$ (See Code 4.3). For the beam search, there are $O(n)$ elements to examine for each of the p retained at a given level. Since the partial tree is examined up to level k, the complexity is in $O(nkp)$ for each element to be added. Since there are n elements to add, the global complexity is in $O(kpn^2)$.

For the pilot method, there is $O(n^3)$ work to do before including an element. The global complexity is therefore in $O(n^4)$, which is confirmed by numerical experiments (see Fig. 11.4).

4.6 Beam Search and Pilot Method Applications

For this problem instance, both the beam search and the pilot methods produce the solution $1 \rightarrow 2 \rightarrow 5 \rightarrow 3 \rightarrow 4 \rightarrow 1$ of length 43. Figure 7 provides the partial solutions successively built for the beam search.

Figure 8 provides the solutions built with the pilot method.

4.7 Greedy Algorithm Implementation for Scheduling

The simplest way is to calculate the increase of the production time if the next object is included at the end (or the beginning) of a sequence. The object with the lowest increase is selected.

A more complex method is to try to insert the next object at all possible positions in the partial sequence. This heuristic, called NEH, is thoroughly studied in the literature.

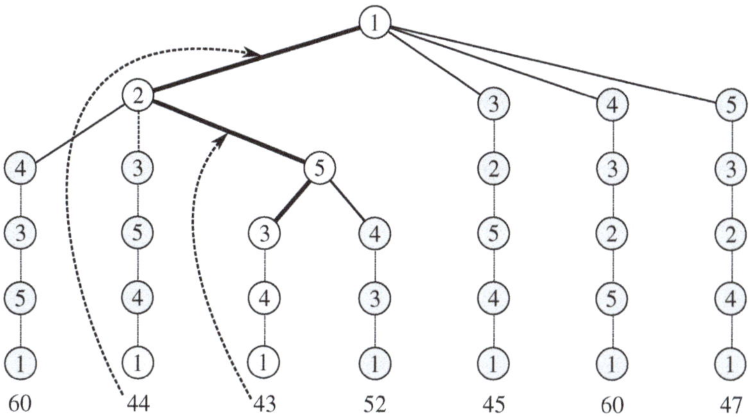

Fig. 8 Applying the pilot method to a TSP instance. The pilot heuristic is the nearest neighbor. The partial solutions completed by the pilot heuristic are drawn with dotted lines

4.8 Greedy Methods for the VRP

A greedy method is given by Algorithm 1.

Create n tours warehouse $\rightarrow i \rightarrow$ warehouse (warehouse $\equiv$ city 0)
forall $i, j = 1, \ldots, n$ **do** Compute savings
  $s_{ij} = c_{i0} + c_{0j} - c_{ij}$ savings that can be achieved by merging tours i and j
Sort the s_{ij} by decreasing value
forall *i, j, in the order of the s_{ij}* **do**
  if *i is at the end of a tour and j at the beginning of another one and the sum of the demands of both tours $\leqslant$ vehicle capacity* **then** Merge the tours
    Add the arc $i \rightarrow j$ and remove the arcs $i \rightarrow 0$ and $0 \rightarrow j$

Algorithm 1 Clarke and Wright's savings greedy algorithm is often cited for building a solution for the vehicle routing problem or for the traveling salesman problem

One can also choose any greedy heuristic for the TSP by initiating it with the depot. As long as there is space left in the vehicle, customers are added to the tour under construction. Then start again with a new vehicle.

Problems of Chap. 5

5.1 Local Minima

See Fig. 9.

The neighborhood has the property of connectivity. x can be increased by 1 unit with the compound move $4 - 3$. x can be decreased by 1 unit with the compound move $4 - 3 - 3 + 4 - 3$.

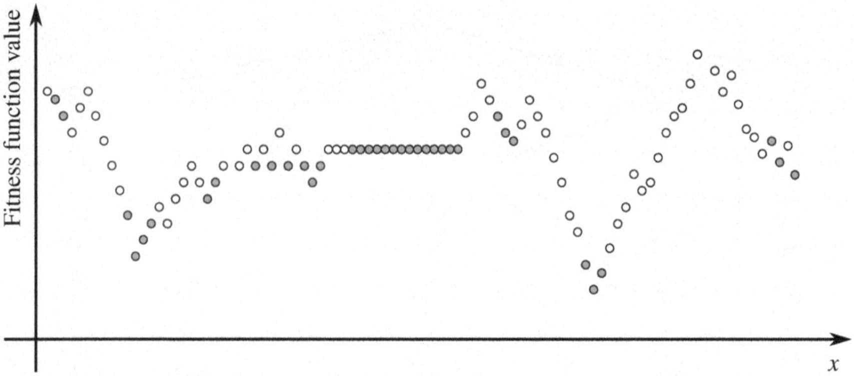

Fig. 9 Local minima of a function of a discrete variable x relative to a neighborhood consisting in either adding 4 or subtracting 3 from x

5.2 Minimizing an Explicit Function

With the first improvement move policy, the following sequences of values are obtained:

$650 \rightarrow 585 \rightarrow 495 \rightarrow 428 \rightarrow 372 \rightarrow 336 \rightarrow 304 \rightarrow 256 \rightarrow 234 \rightarrow 230 \rightarrow 222 \rightarrow 210 \rightarrow 157 \rightarrow 126 \rightarrow 116 \rightarrow 97 \rightarrow 85 \rightarrow 70 \rightarrow 58 \rightarrow 34 \rightarrow 17 \rightarrow 14 \rightarrow 5 \rightarrow -4$

$510 \rightarrow 472 \rightarrow 457 \rightarrow 435 \rightarrow 377 \rightarrow 368 \rightarrow 278 \rightarrow 212 \rightarrow 156 \rightarrow 118 \rightarrow 89 \rightarrow 83 \rightarrow 74 \rightarrow 65 \rightarrow 57 \rightarrow 38 \rightarrow 33 \rightarrow 25 \rightarrow 14 \rightarrow 5 \rightarrow -4.$

With the best improvement move policy, the following sequences of values are obtained:

$248 \rightarrow 193 \rightarrow 138 \rightarrow 123 \rightarrow 89 \rightarrow 58 \rightarrow 34 \rightarrow 17 \rightarrow 14 \rightarrow 5 \rightarrow -10$

$92 \rightarrow 58 \rightarrow 35 \rightarrow 32 \rightarrow 17 \rightarrow -4$

5.3 2-opt and 3-opt Neighborhood Properties

The 2-opt move (i, j) changes the tour

$$i \rightarrow s_i \rightsquigarrow j \rightarrow s_j \rightsquigarrow i$$

to tour:

$$i \rightarrow j \rightsquigarrow s_i \rightarrow s_j \rightsquigarrow i$$

If j is the direct successor of s_i, we can therefore swap two adjacent cities with a 2-opt move. We deduce that the 2-opt neighborhood possesses the connectivity property since we can sort any array with a sequence of adjacent swaps.

The 3-opt move (i, j, k) changes the tour

$$i \to s_i \rightsquigarrow j \to s_j \rightsquigarrow k \to s_k \rightsquigarrow i$$

to tour:

$$i \to s_j \rightsquigarrow k \to s_i \rightsquigarrow j \to s_k \rightsquigarrow i$$

By successively applying the 2-opt moves (i, j), (k, i), (k, s_i), the 3-opt move (i, j, k) is achieved:

$$i \to j \rightsquigarrow s_i \to s_j \rightsquigarrow k \to s_k \rightsquigarrow i$$

$$k \to i \rightsquigarrow s_k \to j \rightsquigarrow s_i \to s_j \rightsquigarrow k$$

$$k \to s_i \rightsquigarrow j \to s_k \rightsquigarrow i \to s_j \rightsquigarrow k$$

With a 2-opt move, one can place any city after any other. Therefore, one can transform any permutation into any other in $n - 1$ steps at most. A 3-opt move allows each city to be moved individually to any place. Thus, the 2-opt and 3-opt neighborhoods have a diameter smaller than n.

5.4 3-opt for Symmetric TSP
There are four possibilities:

$$i \to s_j \rightsquigarrow k \to s_i \rightsquigarrow j \to s_k \rightsquigarrow i$$

$$i \to s_j \rightsquigarrow k \to j \rightsquigarrow s_i \to s_k \rightsquigarrow i$$

$$i \to j \rightsquigarrow s_i \to k \rightsquigarrow s_j \to s_k \rightsquigarrow i$$

$$i \to k \rightsquigarrow s_j \to s_i \rightsquigarrow j \to s_k \rightsquigarrow i$$

Only the first possibility respects the direction of travel of the three sub-paths.

5.5 4- and 5-opt
For 4-opt, there is only one possibility respecting the travel direction. The 4-opt move (i, j, k, u) changes the tour

$$i \to s_i \rightsquigarrow j \to s_j \rightsquigarrow k \to s_k \rightsquigarrow u \to s_u \rightsquigarrow i$$

to tour:

$$i \to s_k \rightsquigarrow u \to s_j \rightsquigarrow k \to s_i \rightsquigarrow j \to s_u \rightsquigarrow i$$

This move is also called *double-bridge*. For 5-opt, there are eight different possibilities respecting the direction of travel of the five sub-paths.

5.6 Comparing 2-opt Best and First

To verify that a solution is 2-optimal, we must test $n(n-1)/2-2$ moves. The number of repetitions of the `while` loop in Code 5.4 grows very slowly with the size of the problem (empirically between $n^{0.1}$ and $n^{0.17}$) because this procedure removes almost all crossings on the first pass.

For Code 5.1, this number of repetitions is almost linear (proportional to $n^{1.1}$). As only one move is performed at each iteration, it can be predicted that, on average, each node is involved a constant number of times in a move during the optimization process.

This increase is virtually independent of the starting solution, but the absolute number of repetitions is about 11 times higher when starting from a random solution than when starting from a solution constructed with a greedy algorithm.

5.7 3-opt Candidate List

A 3-opt move is defined by a triplet (i, j, k). If j and k are limited to 40 values, we can evaluate this limited neighborhood in $O(n)$. Indeed, for each i, there are at most $40 \cdot 39$ neighbors to evaluate. To be able to evaluate this neighborhood, it is necessary to ensure that the city j is indeed on the path $i \rightsquigarrow k$ and not on the path $k \rightsquigarrow i$. It is thus necessary to have a data structure which can supply this information in constant time. This can be the respective position of each city in the tour.

This limited neighborhood is no longer connected.

5.8 VRP Neighborhoods

It is not elementary to construct a feasible solution with a specified number m of tours due to capacity limitations. It is indeed a bin packing problem which is NP-hard (generalization of the set bipartition problem). We can introduce a dummy tour, of unlimited capacity, corresponding to the customers not served by the m ordinary tours. A penalty is associated with each customer on this tour (e.g., the distance depot—customer—depot).

Here are some neighborhoods with the connectivity property for the relaxed problem:

- Take client i and optimally insert it into tour k (dummy or not). Neighborhood size: $O(nm)$.
- Swap customers i and j belonging to different tours. Size of the neighborhood: $O(n^2)$.
- Exchange the beginning of a tour (up to customer i) and the beginning of another one (up to customer j). Neighborhood size: $O(n^2)$.
- Exchange a portion of one tour (between customers i and j) with a portion of another (between customers r and s). Neighborhood size $O(n^4)$ (or $O(m^2)$ if we assume that the number of customers per tour is bounded by a constant).

5.9 Steiner Tree Neighborhood

Model with Steiner vertices to be retained:

- Introduce a new Steiner vertex or delete a Steiner vertex.
- Introduce a new vertex and delete one simultaneously.

This second neighborhood is not connected. Computing the value of a neighboring solution: apply a minimum spanning tree algorithm. The complexity is $O(m + n \log n)$, where m is the number of edges and n is the number of vertices.

Model with connected graph containing incident edges to all terminal vertices: Introducing a new edge (and deleting one if a cycle is created) or deleting one edge (and introducing another if the terminal vertices are no longer in the same connected component). This also implies deleting other edges if there is a connected component solely composed of Steiner vertices. The computation of the solution value is performed using a graph exploration algorithm which is in $O(m + n)$.

5.10 Ejection Chain for the VRP

The chain is initiated by the ejection of a customer i from a tour. The reference structure is a set of tours + an isolated customer. To try a new solution, one attempts to insert i into a tour with sufficient capacity to accept it. The chain can be propagated by inserting i into another tour k while simultaneously ejecting a customer j not yet ejected from that tour. The modified tour k must be feasible and satisfy the capacity constraints. For the propagation, j replaces i. The ejection chain ends if:

- There is no more possible ejection (all the vertices (or a maximum number) were ejected).
- No tour has sufficient capacity to accept i even after deleting j.
- A tried solution is retained.

The complexity of an ejection chain can be established as follows (assuming one goes directly from the preceding customer to the one following the one being ejected, and inserting a customer at the best possible place in the tour). During chain propagation, we try inserting i into the remaining $m - 1$ tours, testing for each insertion all candidates for ejection. This can be done in $O(n)$. The maximum length of the chain is also in $O(n)$. Therefore, a chain can be evaluated in $O(n^2)$. Since there are n different ways to initiate a chain, the overall complexity is in $O(n^3)$.

Problems of Chap. 6

6.1 Dichotomic Search Complexity

The dichotomic search in a sorted array proceeds by dividing the array into $b = 2$ parts. Only one part ($a = 1$) has to be processed recursively. For this problem, there is no reconstruction and the computational effort to be made between two recursive calls is constant ($f(n) = \Theta(1)$).

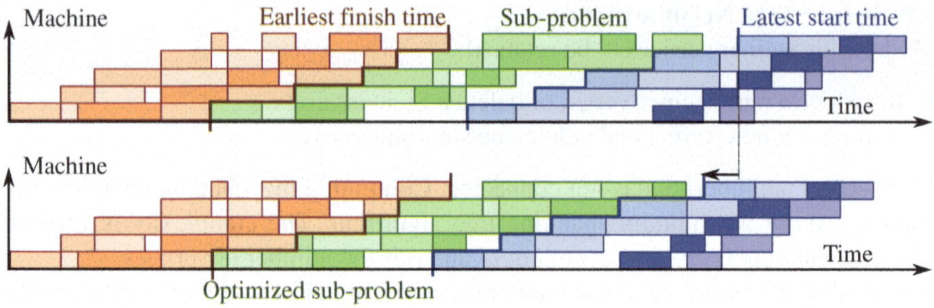

Fig. 10 POPMUSIC for the flowshop scheduling problem: the sequences of objects preceding and succeeding those defining the sub-problem are not changed

Referring to the second case, we find that $T(n) = \Theta(n^{log_2 1} \cdot log\ n) = \Theta(log\ n)$.

6.2 POPMUSIC for the Flowshop Sequencing Problem

In the context of the permutation flowshop scheduling, an object can represent a part for POPMUSIC. A sub-problem consists of the r contiguous objects in the sequence. A sub-problem is optimized with constraints on the earliest start and latest finish times (see Fig. 10).

6.3 Algorithmic Complexity of POPMUSIC

The most complex part of implementing a POPMUSIC method is obtaining an appropriate initial solution. The structure of the initial solution is critical for the method to provide good solutions. It is important to be able to produce this solution with an algorithmic complexity as low as possible. If these conditions are fulfilled, the most significant contribution to the complexity of the framework is the identification of the r parts that make up a sub-problem. If the computational effort to identify a sub-problem depends on the size of the problem, the empirical complexity of the framework is no longer linear.

6.4 Minimizing POPMUSIC Complexity for the TSP

The complexity of building a sample tour is $O(n^{ah})$. The complexity of building the initial tour containing all the cities is $O(n^{h+1})$. The complexity of the optimization with POPMUSIC is $O(n^{1+a-ah})$. For $a \leqslant 1 + \sqrt{2}$, the minimum complexity is reached for $h^* = \frac{a}{a+1}$; for this value, the global complexity is $O(n^{\frac{2a+1}{a+1}})$. This is the typical situation for a first improvement local search with 2-opt neighborhood.

For $a \geqslant 1 + \sqrt{2}$, the minimum complexity is reached for $h^* = \frac{1+a}{2a}$; for this value, the global complexity is $O(n^{\frac{1+a}{2}})$. This is the typical situation for a local search based on Lin-Kernighan neighborhood.

Figure 11 illustrates the complexity of each step of the method as a function of h.

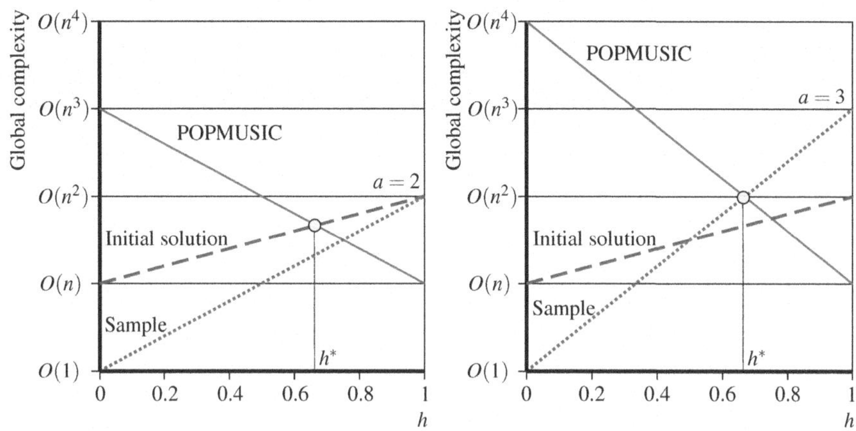

Fig. 11 Diagram used for determining the lowest possible algorithmic complexity as a function of h. Left, when $a = 2$; right, when $a = 3$

Problems of Chap. 7

7.1 SA Duration

A simulated annealing starting with an initial temperature T_0 and ending when the temperature reaches T_f performs $\frac{\log T_f - \log T_0}{\log \alpha}$ iterations if the temperature is multiplied by α at each iteration.

7.2 Tuning GRASP

For this problem instance, the parameter α has almost no influence! Since the starting city is randomly selected in the greedy construction, the latter produces varied initial solutions, even if $\alpha = 0$. The local search Code 12.3 used in this function produces relatively good quality solutions, even if the starting solution is extremely bad. With a less efficient local search, for example, Code 5.4, it is better to choose α close to 1.

7.3 VNS with a Single Neighborhood

When a single neighborhood M_1 is available, a convenient way to implement a variable neighborhood search is to consider that a random move in M_k corresponds to k random moves in M_1.

7.4 Record to Record

The variable neighborhood search implementation can be improved by performing two random swaps at each iteration rather than an increasing number of moves if the solution has not been improved. Code 1 implements such a method.

Code 1 tsp_record_to_record.py Implementation of a record-to-record method. The solution is perturbed by performing two random swaps in the best solution achieved. The method for repairing a perturbed solution is an ejection chain. The method proposed by Dueck [1] includes an additional parameter: a tolerance value of a possible degradation of the solution obtained after the local search. The code provided here would therefore correspond to zero tolerance

```
from random_generators import unif                                # Listing 12.1
from tsp_utilities import tsp_length                              # Listing 12.2
from tsp_LK import tsp_LK                                         # Listing 12.3

######### Record to record iterative local search for the TSP
def tsp_record_to_record(d,                                     # Distance matrix
                         iterations,                         # Number of iterations
                         best_tour,                                    # TSP tour
                         best_length):
    n = len(d[0])
    for iteration in range(0, iterations):
        tour = best_tour[:]          # No tolerance: always revert to best tour

        for _ in range(2):                                 # Perturbate solution
            u = unif(0, n - 1)
            v = unif(0, n - 1)
            tour[u], tour[v] = tour[v], tour[u]
        length = tsp_length(d, tour)
        tour, length = tsp_LK(d, tour, length)
        iteration += 1
        if length < best_length:
            best_tour = tour[:]
            best_length = length
            print('Record to record {:d}\t {:d}'
                  .format(iteration, length))
    return best_tour, best_length
```

Problems of Chap. 8

8.1 Artificial Ants for Steiner Tree

The trails can be stored in an array indexed by the elements of a solution. If we choose a model where an element is a Steiner node, the a priori interest could be the cost of the minimum weight spanning tree over the terminal nodes plus the Steiner nodes selected by the ant. However, this modeling poses a problem: how to decide when the ant should stop incorporating Steiner nodes before returning its solution?

If we select a model where an element e of a solution represents an edge of the tree, the a priori interest is simply the weight of the edge e, if the latter can be added without creating a cycle. The a posteriori interest is proportional to τ_e. An ant builds a solution edge by edge, taking care not to produce a cycle. It can stop as soon as all the terminal nodes are present in the tree. This second model seems to be better adapted to an ant algorithm.

8.2 Tuning the FANT Parameter

For small problem instances, it is challenging to adjust the parameter of the FANT method. Indeed, the local search produces solutions whose quality is almost independent from that of the initial solutions. For numbers of iterations above 100, a parameter τ_b at least equal to 200 seems to produce solutions of moderately better quality.

8.3 Vocabulary Building for Graph Coloring
The solution fragments to be stored in the dictionary can consist of maximum stable sets of the graph. Indeed, all vertices of a stable set can be colored with the same color. A solution can be obtained by selecting a minimum number of stable sets covering all the vertices of the graph. If such a subset of stables can be found, it can be matched with a feasible coloring: a vertex occurring in several of the selected stables will receive an arbitrary color corresponding to one of the stables of which it is a part. In practice, to attempt to obtain a coloring with a fixed number of colors, one constructs a tentative solution with slightly fewer stable sets than this number. The uncovered vertices define a subgraph which is colored independently. If this subgraph is not too large, an exact method can be used. The vertices receiving the same color in this subgraph define a stable set not necessarily maximal in the complete graph. This set can be completed in a maximal stable set and join the solution fragments in the dictionary.

Problems of Chap. 9

9.1 Taboo Search for an Explicit Function
Starting from the solution $(-7, -6)$, the sequence of visited values with a taboo duration $d = 3$ is: $92 \rightarrow 58 \rightarrow 35 \rightarrow 32 \rightarrow 17 \rightarrow -4 \rightarrow -3 \rightarrow -10 \rightarrow 0 \rightarrow -4 \rightarrow 3 \rightarrow 29 \rightarrow 38 \rightarrow 57 \rightarrow 74 \rightarrow 98 \rightarrow 129 \rightarrow 109 \rightarrow 105 \rightarrow 97 \rightarrow 70 \rightarrow 46 \rightarrow 25 \rightarrow -4 \rightarrow 0 \rightarrow -10$.

With a taboo duration $d = 1$, starting from the solution $(-7, 7)$, we have $248 \rightarrow 193 \rightarrow 138 \rightarrow 123 \rightarrow 89 \rightarrow 58 \rightarrow 34 \rightarrow 17 \rightarrow 14 \rightarrow 5 \rightarrow -10 \rightarrow -3 \rightarrow -4 \rightarrow 6 \rightarrow 7 \rightarrow 15 \rightarrow 14 \rightarrow 6 \rightarrow -4 \rightarrow -3 \rightarrow -10 \rightarrow 0 \rightarrow -4 \rightarrow 3 \rightarrow 8 \rightarrow 0$

9.2 Taboo Search for the VRP
Here are some possibilities for defining taboo conditions for the VRP:

- Forbid for d_1 iterations to delete an arc that has just been added and for d_2 iterations to add an arc that has just been deleted.
- Forbid during d iterations to move a customer in a tour that it has left.
- Forbid during d iterations to modify a tour again.

9.3 Taboo Search for the QAP
The solutions successively visited are:

$(1, 2, 3, 4, 5) \rightarrow (2, 1, 3, 4, 5) \rightarrow (3, 1, 2, 4, 5) \rightarrow (3, 2, 1, 4, 5) \rightarrow (2, 3, 1, 4, 5) \rightarrow (4, 3, 1, 2, 5) \rightarrow (4, 3, 5, 2, 1) \rightarrow (4, 2, 5, 3, 1) \rightarrow (2, 4, 5, 3, 1) \rightarrow (2, 4, 5, 1, 3) \rightarrow (3, 4, 5, 1, 2)$.

Table 1 Move evaluation for the application of a taboo search to a small quadratic assignment instance

Iteration	1	2	3	4	5	6	7	8	9	10
Cost	66	58	58	60	60	62	52	52	60	50
Move	Value, **Selected**, *Taboo*									
(1,2)	**−8**	*8*	8	**0**	*0*	8	8	**8**	*−8*	*12*
(1,3)	−6	**0**	*0*	*6*	*6*	4	14	14	4	16
(1,4)	16	14	14	2	**2**	*−2*	*22*	*22*	*0*	22
(1,5)	12	12	12	8	8	8	8	8	12	**0**
(2,3)	0	2	**2**	*−2*	*−2*	*10*	8	8	10	10
(2,4)	2	12	12	10	10	0	**0**	*0*	*14*	12
(2,5)	16	24	24	12	12	12	24	24	12	24
(3,4)	0	20	20	20	20	20	20	20	20	20
(3,5)	4	4	4	14	14	**−10**	*10*	*10*	*10*	20
(4,5)	2	2	2	2	2	0	10	10	**−10**	*10*

Table 1 provides, for the first ten iterations:

- The cost of the solution before performing a move
- The cost differential of each move, with the value of the chosen move in **bold** and in *italics* if it is forbidden

Problems of Chap. 10

10.1 Genetic Algorithm for a One-Dimensional Function

A standard binary representation of a solution is not appropriate. Indeed, two solutions with very close values can have very different representations. For example, (10000000) is completely different from (01111111), even if its value differs by only one unit. For this type of problem, it is better to choose a Gray coding, where the code of x is given by $x \oplus \frac{x}{2}$ ($\oplus$ denoting the bitwise exclusive OR).

10.2 Inversion Sequence

The inversion sequence $(4, 2, 3, 0, 1, 0)$ corresponds to the permutation $(4, 6, 2, 5, 1, 3)$. $(0, 0, 3, 1, 2, 0)$ is not an inversion sequence. Element 5 cannot have two elements greater than itself in a permutation of 6 elements. A sequence $s_i (i = 1, \dots, n)$ is an inversion sequence if and only if $0 \leqslant s_i \leqslant n - i$. The standard crossover operators can be used directly with inversion sequences, as they preserve the property stated above. The drawback is that the corresponding offspring permutations cannot be constructed in linear time.

Table 2 Set of solutions generated in the first iteration of scatter search. The best three become the new elite set, identical ones are deleted. The best two that are the most different from an elite solution are retained

Generated solution	Value	Volume	Repaired solution	Value	Volume	Rank	Distances
(0 0 1 1 0 0 0 0 0 0)	21	30	(0 1 1 1 0 0 0 1 0 0)	36	90	7	3 5 3
(0 1 1 1 1 0 0 0 1 0)	44	100	(0 1 1 1 1 0 0 0 1 0)	44	100	1	Elite
(0 1 1 1 0 0 0 0 1 1)	42	89	(0 1 1 1 0 0 0 0 1 1)	42	89	3	Elite
(0 1 1 1 0 0 0 0 1 1)	42	89	(0 1 1 1 0 0 0 0 1 1)	42	89	Deleted	
(0 1 1 1 1 0 0 0 1 0)	44	100	(0 1 1 1 1 0 0 0 1 0)	44	100	Deleted	
(1 0 1 1 1 0 0 0 0 0)	42	92	(1 0 1 1 1 0 0 0 0 0)	42	92	4	4 3 5, Retained
(1 0 1 1 1 0 0 0 0 0)	42	92	(1 0 1 1 1 0 0 0 0 0)	42	92	Deleted	
(0 1 1 1 1 0 0 0 1 0)	44	100	(0 1 1 1 1 0 0 0 1 0)	44	100	Deleted	
(0 1 1 1 1 0 0 0 1 0)	44	100	(0 1 1 1 1 0 0 0 1 0)	44	100	Deleted	
(1 0 0 1 0 0 1 0 0 1)	38	96	(1 0 0 1 0 0 1 0 0 1)	38	96	6	6 3 5
(0 1 1 1 1 0 0 0 1 0)	44	100	(0 1 1 1 1 0 0 0 1 0)	44	100	Deleted	
(1 0 1 1 0 0 0 0 0 1)	40	81	(1 0 1 1 0 0 0 0 1 1)	43	95	2	Elite
(0 1 1 1 0 0 0 0 0 1)	39	75	(0 1 1 1 0 0 0 0 1 1)	42	89	Deleted	
(0 1 1 1 0 0 0 0 1 1)	42	89	(0 1 1 1 0 0 0 0 1 1)	42	89	Deleted	
(0 1 1 1 0 0 0 0 1 1)	42	89	(0 1 1 1 0 0 0 0 1 1)	42	89	Deleted	
(0 1 0 1 0 0 0 0 0 1)	30	59	(0 1 0 1 0 0 0 1 0 1)	35	92	8	5 5 3
(1 0 1 1 1 0 0 0 0 0)	42	92	(1 0 1 1 1 0 0 0 0 0)	42	92	Deleted	
(0 1 1 1 1 0 0 0 0 0)	41	86	(0 1 1 1 1 0 0 0 1 0)	44	100	Deleted	
(1 0 0 1 0 0 0 0 0 1)	31	65	(1 1 0 1 0 0 0 0 0 1)	41	92	5	5 3 3, Retained
(0 1 0 1 0 0 0 0 0 1)	30	59	(0 1 0 1 0 0 0 1 0 1)	35	92	Deleted	
(0 1 1 1 1 0 0 0 1 0)	44	100	(0 1 1 1 1 0 0 0 1 0)	44	100	Deleted	
(0 1 1 1 1 0 0 0 1 0)	44	100	(0 1 1 1 1 0 0 0 1 0)	44	100	Deleted	
(1 0 1 1 0 0 0 0 0 1)	40	81	(1 0 1 1 0 0 0 0 1 1)	43	95	Deleted	
(0 1 1 1 0 0 0 0 1 1)	42	89	(0 1 1 1 0 0 0 0 1 1)	42	89	Deleted	
(0 0 1 1 1 0 0 0 0 0)	31	59	(0 1 1 1 1 0 0 0 1 0)	44	100	Deleted	
(0 1 1 1 0 0 0 0 0 1)	39	75	(0 1 1 1 0 0 0 0 1 1)	42	89	Deleted	

In the context of scatter search, k solutions $s^1, \dots, s^k$ can be mixed by rounding the elements of $\frac{\sum f(s^i)s^i}{\sum f(s^i)}$ (with $f(\cdot)$ to maximize).

10.3 Rank-Based Selection

The probability of `rank_based_selection(m)` to return v is $\frac{2\cdot(m-v+1)}{m\cdot(m+1)}$.

10.4 Tuning a Genetic Algorithm

Two parameter settings seem to be appropriate: with a zero mutation rate, a relatively large population (100 solutions) should be adopted. With a single random mutation after crossover, a population of a dozen solutions is adequate.

10.5 Scatter Search for the Knapsack Problem

Table 2 provides the list of solutions produced at the first generation of a scatter search applied to a knapsack instance.

Table 3 Empirical number of elements found by the Pareto local search and number of elements compared when using a KD-tree or a linked list. n is the number of TSP cities. For $K = 4$ and $n = 30$, the number of comparisons is larger than $3 \cdot 10^{11}$ with a KD-tree. With a linked list, it would have taken several weeks or months of calculation for the program to finish

	Pareto size		Comparisons	
K	$n = 30$	grows	KD-tree	list
2	82	$\overline{O}(n^{1.5})$	$\overline{O}(n^{6.5})$	$\overline{O}(n^{7.5})$
3	6970	$\overline{O}(n^{3.7})$	$\overline{O}(n^{8.2})$	$\overline{O}(n^{9.4})$
4	548671	$\overline{O}(n^{6.1})$	$\overline{O}(n^{10.6})$	$\overline{O}(n^{12.6})$

Problems of Chap. 11

11.1 Data Structure Options for Multi-objective Optimization
The number of comparisons grows much faster with a linked list than with a KD-tree. This growth seems polynomial with the number of cities in the problem instance. The size of the Pareto set also grows polynomially. The degree of these polynomials grows very strongly with K, the number of objectives. Table 3 gives an estimate of these degrees for instances with random distance matrices, uncorrelated between objectives. Because of the recursive algorithm for deleting an element in a KD tree, about twice as many element removal queries must be made as with a list.

11.2 Comparison of a true Simulated Annealing and a Kind of SA with Systematic Neighborhood Evaluation
The noising method code that systematically evaluates the neighborhood allows many more iterations for the same computation time. At low temperatures, convergence is faster, as shown in Fig. 12. Both methods stop after just under 10 million iterations.

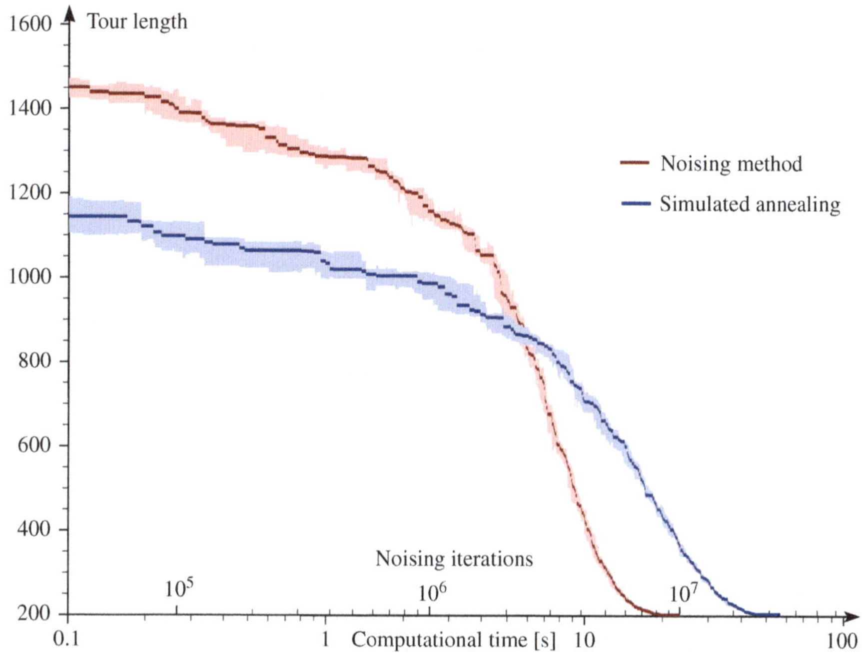

Fig. 12 Comparison of a basic simulated annealing and a simulated annealing with systematic neighborhood evaluation (kind of noising method). Median solution value as a function of the number of iterations

Reference

1. Dueck G.: New optimization heuristics: The Great Deluge Algorithm and the record-to-record travel. J. Comput. Phys. **104**(1), 86–92 (1993) https://doi.org/10.1006/jcph.1993.1010

Index

É. D. Taillard, *Design of Heuristic Algorithms for Hard Optimization*, Graduate Texts in Operations Research, https://doi.org/10.1007/978-3-031-13714-3

The manufacturer's authorised representative in the EU is Springer Nature Customer Service Centre GmbH, Europaplatz 3, 69115 Heidelberg, Germany. If you have any concerns regarding our products, please contact ProductSafety@springernature.com

Printed and bound by CPI Group (UK) Ltd, Croydon, CR0 4YY
22/09/2026
02221055-0013